T0073479

Water and the Future of Humanity

Gulbenkian Think Tank on Water and the Future of Humanity

Water and the Future of Humanity

Revisiting Water Security

The Gulbenkian Think Thank on Water and the Future of Humanity

Benedito Braga, Colin Chartres, William J. Cosgrove, Luis Veiga da Cunha, Peter H. Gleick, Pavel Kabat, Mohamed Ait Kadi, Daniel P. Loucks, Jan Lundqvist, Sunita Narain, Jun Xia (in alphabetic order)

CALOUSTE
GULBENKIAN
FOUNDATION

 Springer

Gulbenkian Think Tank on Water
   and the Future of Humanity
Calouste Gulbenkian Foundation
Avenida de Berna 45A
1067-001 Lisbon, Portugal

ISBN 978-3-319-01456-2   ISBN 978-3-319-01457-9 (eBook)
DOI 10.1007/978-3-319-01457-9
Springer New York Heidelberg Dordrecht London

Library of Congress Control Number: 2013946034

© Calouste Gulbenkian Foundation 2014
This work is subject to copyright. All rights are reserved by the Publisher, whether the whole or part of the material is concerned, specifically the rights of translation, reprinting, reuse of illustrations, recitation, broadcasting, reproduction on microfilms or in any other physical way, and transmission or information storage and retrieval, electronic adaptation, computer software, or by similar or dissimilar methodology now known or hereafter developed. Exempted from this legal reservation are brief excerpts in connection with reviews or scholarly analysis or material supplied specifically for the purpose of being entered and executed on a computer system, for exclusive use by the purchaser of the work. Duplication of this publication or parts thereof is permitted only under the provisions of the Copyright Law of the Publisher's location, in its current version, and permission for use must always be obtained from Springer. Permissions for use may be obtained through RightsLink at the Copyright Clearance Center. Violations are liable to prosecution under the respective Copyright Law.
The use of general descriptive names, registered names, trademarks, service marks, etc. in this publication does not imply, even in the absence of a specific statement, that such names are exempt from the relevant protective laws and regulations and therefore free for general use.
While the advice and information in this book are believed to be true and accurate at the date of publication, neither the authors nor the editors nor the publisher can accept any legal responsibility for any errors or omissions that may be made. The publisher makes no warranty, express or implied, with respect to the material contained herein.

Printed on acid-free paper

Springer is part of Springer Science+Business Media (www.springer.com)

Foreword

Improved water supply and sanitation contributes to human dignity, health, education, and economic development. It is also increasingly clear that children and adults suffering from diarrhea, as a result of unsafe water and lack of sanitation facilities, cannot fully absorb the nutrients they consume, a condition often referred to as environmental enteropathy. Water has no substitute and the demands for it are increasing. Food and energy production along with industrial activity increasingly compete for the same water to meet the needs of a growing world population. At the same time, we are now seeing the impacts of climate change on water resources management.

This book is the result of 2 years of fruitful discussion about water and the future of humanity among 11 of the world's experts in water resources management under the sponsorship of the Calouste Gulbenkian Foundation. They have examined the major challenges that we face today and that our children will face in the future. They describe the use of water by sector and geographic area, the benefits for humans, and the way in which water use impacts other sectors. They remind us that the source of all our water is precipitation and that water resources are continuously replenished through the hydrological cycle—a cycle driven by energy from the sun. Natural and man-made ecosystems, of which humans are a part, play a vital role in water resources dynamics. The fragility of ecosystems to over-abstraction and pollution is highlighted.

This is a book not only about water management challenges but also about water's value across multiple sectors. We fail to recognize basic facts. The truth is that often water is undervalued in many of the products and services it provides. It is obvious that we must urgently manage our water use more efficiently, even so, under optimistic future scenarios; there could still be millions excluded from access to water and the goods and services that water provides. Wealthy countries and societies can afford expensive technology or purchase what they need on the world market. Countries and people trapped by poverty and suffering as a result of climate change may not have this luxury.

In describing the challenges, the authors send a message of hope. There will almost certainly be advances in technology and know-how in the years to come. They describe technological, economic, and institutional advances already being implemented by decision makers in different settings throughout the world. They conclude that by replicating and scaling up these advances we can overcome the challenges we face to create a water-secure world. Leadership, vision along with individual and collective action will help secure the future we want.

December 2012

HRH the Prince of Orange, Chair, UN Secretary General's Advisory Board on Water Supply and Sanitation

Prologue

The Calouste Gulbenkian Foundation has a long history of taking an interest in global issues that affect the world today and can, eventually, create serious problems and constraints to the development of humanity. More recently, it has given due consideration to the analysis of environmental problems and processes. The convergence of these two concerns led to the creation of the "Gulbenkian Think Tank on Water and the Future of the Humanity," initiated in 2010, in the context of the former Gulbenkian Environment Program.

One of the purposes of the Think Tank was to reflect on the possibility of future water use creating serious barriers to human development. This possibility has not received sufficient attention so far and, consequently, the issue is not yet a source of adequate public concern.

The Foundation invited 11 highly respected personalities from the science and water management communities to be part of the Think Tank. They have been chosen on the strength of their sound knowledge and experience, as well as their broad vision of freshwater issues and problems. The members of the Think Tank come from different areas of knowledge and different geographic regions. They were all invited on a personal basis, not as representatives of any organizations or institutions they are associated with.

As a result of its activity, the Gulbenkian Think Tank has produced the present book.

The Calouste Gulbenkian Foundation expresses its deep gratitude to the members of the Think Tank for their generous and committed participation in this initiative.

We would also like to thank Luis Veiga da Cunha, coordinator of the group. His vision and expertise on these questions enriched the discussions held on this project since its early stage and were an important element in the Foundation's decision to pursue it.

The mandate of the Think Tank was to clarify the main drivers and issues of an eventual water crisis, in order to identify a realistic vision of a water-secure world in the next few decades and to suggest possible ways to cope with related future water problems.

In the current times of change and uncertainty, global water security is, no doubt, a major concern. Similarly, interactions and feedback effects running through water and other sectors, such as food, energy, urban development, and biodiversity, present major and complex challenges in the globalized world of tomorrow. In this context, water conveys a potential for crisis and conflict, since it lays at the core of most things which are important for human life. Moreover water, although renewable, is limited and has no substitutes.

We are, thus, convinced that this initiative is a timely effort. We believe that the results of the "Think Tank on Water and the Future of Humanity" represent a valuable contribution for ongoing debates on global sustainable development. In particular, they could be helpful to clarify the need for a global water governance, as a cooperative arrangement to ensure improved living standards for the next generations.

Artur Santos Silva
President, Calouste Gulbenkian Foundation

Preface

The future does not belong to anybody.
There are no precursors, only latecomers.

Jean Cocteau

The Calouste Gulbenkian Foundation in Lisbon, Portugal, decided to create an international Think Tank to analyze noticeable features of the dynamic interactions between freshwater systems and society in the twenty-first century. This book is the result of such an initiative. I was honored by the invitation of the Foundation Board of Trustees to advise on the composition and working rules of this Think Tank and, later, to act as coordinator of the Think Tank activity.

The aim of the Gulbenkian Think Tank was to enhance present knowledge on the role of water in the world. Its reflections addressed water use until 2050, as well as the state of water resources in the planetary environment. The Think Tank has also reflected on the possible creation of serious barriers to development, caused by water-related constraints.

The Think Tank has examined likely trends, regarding water availability and management, comparing them with the growing water demand from various sectors. The Think Tank has also analyzed the main driving forces at play, in order to access the kind of human effort that is feasible and desirable to cope with future situations.

The members of the Gulbenkian Think Tank are (in alphabetical order): Prof. Benedito Braga (*President, World Water Council*; *Professor, Escola Politécnica da Universidade de São Paulo, Brazil*); Dr. Colin Chartres (*Director General, International Water Management Institute—IWMI, Colombo, Sri Lanka*); Dr. William J. Cosgrove (*Honorary President World Water Council, Montreal, Canada*); Prof. Luis Veiga da Cunha (*Professor, Universidade Nova de Lisboa, Lisbon, Portugal*); Dr. Peter Gleick (*President, Pacific Institute, Oakland, USA*); Prof. Pavel Kabat (*Director and CEO, International Institute for Applied Systems Analysis—IIASA, Austria*; *Professor of Earth Systems Science, Wageningen University, the Netherlands*); Dr. Mohamed Ait Kadi (*President, Conseil General du Développement Agricole, Rabat, Morocco*); Prof. Daniel P. Loucks (*Professor, Cornell University, Ithaca, USA*); Prof. Jan Lundqvist (*Senior Scientific Advisor, Stockholm International Water Institute—SIWI, Stockholm, Sweden*); Ms. Sunita

Narain (*Director General, Centre for Science and Environment, New Delhi, India*); and Prof. Jun Xia (*Chair Professor and Dean, Research Institute for Water Security, Wuhan University, China*).

The members of the Gulbenkian Think Tank have assumed collective authorship of the whole book. Obviously, the members agreed to initially distribute the preparation of drafts of the book chapters, which were subsequently discussed in depth by all group members. The book is, thus, a true collective work. The book went through three successive editing processes: first, a scientific editing by Professor Daniel P. Loucks, member of the Think Tank; second, a professional editing by the London firm Scriptoria; and third, the final editing undertaken by the publishers of the book.

The global direct and indirect water demand in 2050 was considered, with reference to changes in population and GDP of countries, grouped into seven regions of the world. Global demographic growth, from the current more than seven billion inhabitants of the Earth to more than nine billion by 2050, will be a major driver of changes in water demand. These changes will substantially increase the pressures on water systems towards 2050, with special implications for food, energy, and the environment.

A host of multidimensional drivers is related to an expansion of the so-called *urban culture*, a feature in mushrooming cities. Currently, about half of the world's population lives in urban centers. By 2050, more than 70 % of the world's population will be living in urban areas. The impacts of the corresponding shifts in demand for food, water, and energy will be felt far beyond the boundaries of the urban centers themselves.

Considering increasing demands for water in line with its more uncertain availability, environmental water requirements will become a hot topic. The demand for water has often been considered to be at odds with the need for water to maintain the life of multiple organisms. In addition, human waste, and particularly wastewater, has been discharged into the environment with little concern for its impact on ecosystems. Depriving ecosystems of water, essential to their life, and poisoning them with waste would feedback negatively on human life and development.

The availability of water can become a serious constraint on development. This may happen in a relatively short period. Thus, the danger of insufficient timely awareness is very real. As some have already claimed, a water crisis could negatively affect humanity, even more than the much discussed climate change crisis. However, the public and political concern with global warming is currently stronger than the concern with a global water crisis. The Earth may be quickly approaching critical tipping points related to water, food, and energy security. It is important to recognize that the water and the climate crisis are closely interrelated. Global warming will affect water supply and demand, as well as water quality. At the same time, water is clearly the main mediator of the impacts of climate change in the economy, society, and the environment.

There is a clear need for an interdisciplinary and inter-sectorial reflection on the processes and issues involved in the anticipated global water scarcity and security problems, which may seriously affect the future of humanity.

Concerned with the different aspects referred to above, the Gulbenkian Think Tank has proposed a "Message on Water and the Future of Humanity," which is presented just before the book text.

The book consists of nine chapters. The two initial chapters address, in an introductory manner, a number of issues particularly related to future water problems. These include the relationship between development, environment and water, the increasing water crises in the Anthropocene, water and globalization, and water governance (Chap. 1), and also the drivers of water demand, course changes, envisioning the future, projecting water demands and the need for a change of human behavior, involving modified social and environmental concerns (Chap. 2).

The following six chapters deal in detail with a number of critical factors already present but deemed to increase in the future. They relate to water in a variable and changing climate (Chap. 3); water, the environment and ecosystems services (Chap. 4); water in an urbanizing world (Chap. 5); water and food security (Chap. 6); water and energy nexus (Chap. 7); and water projections and scenarios (Chap. 8).

The book concludes by reviewing the main water-related challenges confronting humanity, followed by the consideration of ways to respond to these challenges, emphasizing the role of leadership, commitment, and responsibility (Chap. 9).

The book is intended to present the current knowledge about the challenges, risks, and opportunities present in our path to a future water sustainable world. Water sustainability is something much too serious to leave to politicians, managers, and scientists alone. It is a crucial issue for our emerging globalized world. It concerns everybody and will strongly condition the quality of our lives and even our survival.

The Gulbenkian Think Tank on Water and the Future of Humanity has aimed to offer a scientifically sound book and, at the same time, a readable and motivating one for the public in general. If the business-as-usual approaches, currently adopted to cope with water problems, are not drastically changed then the future of water and, as a consequence, our own welfare may be seriously threatened. We must take care not to be latecomers to the Future, as mentioned by Cocteau in the opening quote of this Preface.

Luis Veiga da Cunha
Coordinator of the Gulbenkian Think Tank
on Water and the Future of Humanity

Acknowledgments

The book has benefited significantly from contributions and advice from scholars outside the Gulbenkian Think Tank on Water and the Future of Humanity. This was done by so many, and in so many ways, that we cannot thank them all personally, but we particularly want to acknowledge the following:

Dr. David Wiberg, International Institute for Applied Systems Analysis (IIASA), together with Fulco Ludwig, Wageningen University, for extensive contributions to the water and climate change topic; Dr. Aditya Sood, International Water Management Institute (IWMI), for re-implementing the WATERSIM model and for providing scenarios of water demand and water availability for the period 2010–2050; Morgan Levy, University of California, Berkeley, for her work at the Pacific Institute researching and writing on the history and practice of water scenarios and projections; Camelia Dewan, Stockholm International Water Institute (SIWI), and Dr. Gerald C Nelson, International Food Policy Research Institute (IFPRI), for providing data on estimated GDP growth for the period 2010–2050; Dr. Dieter Gerten and Holger Hoff, Institute for Climatic Research (PIK), for providing global hydrological data as an input for the WATERSIM model; Dr. Mark Smith, International Union For Conservation of Nature (IUCN), and Dr. Karin Krchnak, The Nature Conservancy, for their contributions to water and the environment; and Ms. Debra Perrone, Vanderbilt University, for her advice on water and energy issues.

We would like to acknowledge SCRIPTORIA (London) and TVM DESIGNERS (Lisbon), respectively for their professional editing of the book and for redrawing the figures and tables to improve their uniformity in style and quality. Finally we also want to acknowledge with thanks the intellectual and logistic support we received from IIASA, IWMI, and SIWI during the preparation of this book.

Country borders or names shown on maps in this book do not necessarily reflect the position of the Calouste Gulbenkian Foundation or any of the members of the Gulbenkian Think Tank on Water and the Future of Humanity concerning the legal status of any country or territory or concerning the delimitation of frontiers or boundaries.

A Message on Water and the Future of Humanity

A combination of demographic, technological, and economic trends has accelerated the ability of humans to modify their built and natural habitats. Growth in the world's economic sectors—including agriculture—and urban centers is changing the landscapes of our Earth and multiplying the flow of goods and services derived from water and other natural resources. The human potential to develop and manage natural resources for the individual and common good has increased exponentially during recent decades. Unfortunately, many of our past governance and piecemeal decisions concerning the management and use of natural resources have resulted in real and possible threats. These threats take the form of disruption of river flow regimes and water quality, the lowering of ground water tables, and the deterioration of natural ecosystems and of the services they provide, including those related to water quality. In addition, global warming, including climate change and variability, speeds up many of the processes in the hydrological cycle with increased unpredictability in the spatial and temporal availability of water. Geographically, as well as economically and socially, there will be winners and losers if these trends continue unabated.

Being the dominant and most dynamic species on Earth, humans have an opportunity to willfully reflect on the consequences of their collective behavior. We can use our visionary abilities, technologies, and economic resources for increased human well-being and the sound stewardship of our resources. Or we can allow a business-as-usual trend to continue, with its inherent risk for undesirable disruptions on planet Earth and, in particular, for our water resources. Piecemeal and post-damage control of undesirable disruptions in life support systems are not enough in the environment we find ourselves in today; an environment where humans can indeed control the fate of this planet. In this Anthropocene epoch, humans are the only species with the capacity to reflect on their behavior and change it as needed to achieve the goals of humanity in harmony with the dynamics of water and the other natural resources essential for our health and economic and social well-being.

The writers of this book seek to determine a path to a desirable, sustainable future for humans in a world with immense opportunities, but also with limits, boundaries, and vulnerabilities. We believe essential components of human well-being can be enhanced without a corresponding increase in resource exploitation and its associated detrimental environmental side effects. Inefficiencies in the use of water and other resources are still low when compared to the inefficiencies in the use of our productive potential. There are a number of obstacles to overcome. Providing healthy lives and meaningful livelihoods for humankind require a number of changes in the way that the necessary water, food, energy, and other goods and services are provided and beneficially consumed. Changes are also required in the ways in which by-products are recycled or disposed of. Progress has to be envisioned against a background of the present consumption habits of the most affluent, the inequitable distribution of the benefits of the planet's water and other resources, and a climate that is changing because of, among other reasons, human activity.

We have examined in a systematic manner the major development challenges and the linkages between them and their dependence and effects on water and other resources. From this examination we have identified several positive trends and technological, economic, political, and social measures that, if adopted, would set the course to the achievement of a desirable future.

The Gulbenkian Think Tank on Water and the Future of Humanity

Contents

Foreword
by HRH the Prince of Orange, Chair, UN Secretary General's Advisory
Board on Water Supply and Sanitation   v

Prologue
by Artur Santos Silva, President of Calouste Gulbenkian Foundation   vii

Preface
by Luis Veiga da Cunha, Coordinator of the Gulbenkian
Think Tank on Water and the Future of Humanity   ix

Acknowledgements   xiii

Message on Water and the Future of Humanity
by the Gulbenkian Think Tank on Water
and the Future of the Humanity   xv

1   Our Water, Our Future...   1
    1.1   Considering the Future..   1
    1.2   Development, Environment, and Water...   2
    1.3   The Anthropocene and Water..   5
    1.4   The Increasing Water Crisis...   7
    1.5   Sustained Prosperity While Controlling Water Use...........................   10
    1.6   Important Future Water Issues..   12
    1.7   Water and Globalization..   13
    1.8   Water Governance..   14
    1.9   Toward a Global Water Governance?..   17
    1.10 Creating a Water Secure World..   19

2   Drivers of Water Demand, Course Changes, and Outcomes...............   21
    2.1   Staggering Growth, Water Use, and Human Behavior.......................   21
    2.2   Human Ambition and Capacity to Modify Our Planet......................   22

2.3 Visioning Our Future... 23
    2.3.1 Predicting the Future of the Planet... 25
    2.3.2 Feedback Mechanisms.. 27
    2.3.3 Altering the Trajectory of Development Efforts 28
2.4 Lessons from the Past.. 28
    2.4.1 Water Infrastructure as "Temples of Modernization" 30
    2.4.2 Water Provision and Regulation Against Floods 31
    2.4.3 Increasing Dependence on Groundwater................................. 31
2.5 Different Types of Water and Their Relative Significance............... 32
2.6 Checks and Balances: Drivers That Modify Thinking
    and Policy.. 32
2.7 A Convenient Truth: Trend Break in Population Dynamics............. 33
2.8 The World Economic Map Turns Upside Down.............................. 35
    2.8.1 Income Distribution Determines Access to,
         and Mix of, Goods and Services... 36
2.9 Projecting Water Demands.. 36
    2.9.1 Efficiency, Effective Goal Achievement,
         and Net Water Savings... 40
    2.9.2 New Thinking for Opportunities to Tackle Old
         and New Challenges... 43
    2.9.3 Linking Sustainable Production to Fair Access...................... 43
2.10 Prospects for an Alternative "Privileged Problem Formulation"...... 44

3 Water Management in a Variable and Changing Climate................... 47
3.1 Climate Change and Water Management.. 47
3.2 Effects of Climate Change on Water.. 48
    3.2.1 The Earth's Energy Balance and Hydrologic Cycle.............. 48
    3.2.2 Recent Observed Changes in Temperature
         and Precipitation.. 50
    3.2.3 Recent Changes in Water Availability.................................... 50
    3.2.4 Hydrological Extremes... 51
    3.2.5 Future Changes in Water Availability..................................... 52
    3.2.6 Effects of Climate Change on Water Demand....................... 54
3.3 Basics of Climate Science: What a Water Manager
    Needs to Know.. 56
    3.3.1 Climate Change Assessment Process..................................... 56
    3.3.2 Historic Data.. 57
    3.3.3 Scenario Development... 59
    3.3.4 Modeling.. 60
3.4 Uncertainty in Water Management.. 66
    3.4.1 The Dutch Delta: An Example of the Possibilities
         for Floodplain Management... 68
    3.4.2 China: The Challenge of Jointly Managing
         Great Diversity, Rapid Socioeconomic Changes,
         and Climate Adaptations.. 71

        3.4.3 Climate Change Adaptation in Bangladesh 74
  3.5 The Way Forward 76

4 Water for a Healthy Environment 79
  4.1 Ecosystems, the Environment, and Humans 79
        4.1.1 Limits to Withdrawals of Water from Nature 80
        4.1.2 Ecosystem Services 81
        4.1.3 Ecosystem Services and Poverty. 83
        4.1.4 Environmental Flows 84
  4.2 Human Impact on the Environment 85
        4.2.1 Agriculture 86
        4.2.2 Urbanization and Industrialization 87
        4.2.3 Desertification 89
        4.2.4 Coastal Zones 90
        4.2.5 What We Don't Know 91
        4.2.6 Tipping Points 93
  4.3 Changing Approaches to Water Management 94
        4.3.1 Integrated Water Resource Management (IWRM)
              and Ecosystem Management 95
        4.3.2 Win-Win Approaches 96
        4.3.3 Engineered Ecosystems 98
  4.4 Valuing and Allocating Water and Ecosystems 101
        4.4.1 A Green Economy 103
        4.4.2 Trade-Offs 105
  4.5 Who Speaks for Mother Nature? 105
  4.6 Meeting Human Needs on a Sustainable Planet 106
  4.7 Summing It Up 107

5 Integrated Urban Water Resources Management 109
  5.1 Urban Centers and Urban Water Systems 109
        5.1.1 Current Conditions in Urban Centers 111
        5.1.2 Aging Urban Infrastructure in Developed Regions 112
        5.1.3 Current Urban Water Management 112
        5.1.4 Urban Centers of the Future 114
  5.2 The Need for Integrated Urban Water Resources Management 115
        5.2.1 Water Supply and Sanitation 117
        5.2.2 The Role of Technology 119
        5.2.3 Stormwater Drainage 124
        5.2.4 Water and Energy Nexus 127
        5.2.5 Governance 128
  5.3 The Way Ahead 131

**6 Water and Food Security: Growing Uncertainties
and New Opportunities** 133
  6.1 Uncertainty Dominates 133
  6.2 Our Journey Toward Food Security 134

      6.2.1 The Evolution of the Food Security Concept............................ 134
      6.2.2 A New Food Equation.. 139
  6.3 Hunger for Land and Thirst for Water 141
  6.4 The Century of Agriculture.. 146
  6.5 Climate Change Consequences.. 151
  6.6 Smarter Management of Water and Food Systems 153
  6.7 Toward a Viable Food Future.. 156

7 Water and Energy.. 159
  7.1 Understanding the Water–Energy Nexus 159
  7.2 Water for Energy 162
  7.3 Energy for Water 168
  7.4 Analyzing Water–Energy Systems.. 171
  7.5 Efficiency Potentials and Challenges..................................... 174
  7.6 Water, Energy, and Climate Change...................................... 177
  7.7 Water and Energy Security.. 179
  7.8 Water and Energy Governance.. 182
  7.9 Water and Energy: Future Perspectives.................................. 183

8 Water Projections and Scenarios: Thinking About Our Future 185
  8.1 Thinking About Our Water Futures....................................... 185
  8.2 History of Water Scenario Development and Use........................ 186
  8.3 Data Constraints.. 189
  8.4 Positive Water Scenarios.. 191
  8.5 Results for Positive Water Scenarios..................................... 193
  8.6 Backcasting Approaches... 195
  8.7 Backcasting Experience in Water Planning.............................. 196
  8.8 The Concept of "Water Wedges".. 199
  8.9 Soft Water Paths.. 201
  8.10 Strategies for Moving to a Positive Future............................. 203
      8.10.1 Water-Independent Drivers and Strategies......................... 204
      8.10.2 Water-Focused Drivers and Strategies.............................. 204
  8.11 Moving from Here to There: Where Do We Want to Be?................ 205

9 Our Water Future: Leadership and Individual Responsibility........... 207
  9.1 Water Management Challenges.. 207
  9.2 Meeting Human Needs: Water-Related Challenges...................... 208
      9.2.1 Food Security.. 208
      9.2.2 Energy Security.. 209
      9.2.3 Urban and Industrial Demand....................................... 210
      9.2.4 Resource Security.. 211
      9.2.5 Meeting Human Water Demands in 2050............................ 212

9.3 Climate Change and Governance.. 214

9.3.1 Climate Change.. 214

9.3.2 Appropriate and Effective Water Governance 216

9.4 Responding to Water-Related Challenges.. 217

9.5 Leadership, Commitment, and Responsibility..................................... 219

References... 221

**Biographic Notes of the members of the Gulbenkian Think Tank
on Water and the Future of Humanity**... 237

Our Water, Our Future

1

Water plays a unique role on our planet. The multiple challenges related to the management of water are huge. A changing climate and rapidly growing populations, social and economic development, globalization, and urbanization are among the external drivers shaping our current world. Some suggest that soon we will be witnessing a water crisis. Therefore, we must act before such a crisis become inevitable and irreversible. All of us can, and must, find ways to safeguard water and, with it, the future of humanity. We must set in motion new strategies governing the way we live and interact with our environment to ensure that there will be enough water to support, rather than constrain, future generations. Global sustainability is, fundamentally, about the way people live and interact with our Earth. In the Anthropocene in which we live, we also have the ability to influence the future of our freshwater resources and the future of humanity.

1.1 Considering the Future

Understanding the past, describing the present, and analyzing the future reflects a need to have a perspective on who we are and where we are going, and why. Attempting to forecast the future is particularly relevant in this unique period of human history given the multiple signs that suggest we are in the trajectories of increasingly troubled situations. Predicting the future is impossible, if for no other reason than it can depend to some extent on what we humans make of it. Given the increasing pressures on our environment and society, it is critical that we attempt to consider our future in order to actively guide it to what we want it to be rather than just reacting and adapting to what happens. Because predicting the future is impossible, we are forced to follow two courses of action. First, we must consider what kind of future we would like for ourselves and for our descendants, and second, we must decide which actions to take today in order to ensure the best results in the future.

Understanding the present world and its future evolution requires interdisciplinary knowledge. It requires an understanding of each of the drivers of change.

Gulbenkian Think Tank on Water and the Future of Humanity, *Water and the Future of Humanity: Revisiting Water Security*, DOI 10.1007/978-3-319-01457-9_1,
© Calouste Gulbenkian Foundation 2014

Throughout human history, climate and weather have influenced what humans could and could not do. Science and technology have helped us to adapt to our changing environment. Mainly through the burning of fossil fuels and changing land use, we are now witnessing substantial changes in our climate. At the same time, science and technology are constantly providing us with new knowledge and tools. The big question is whether we can use this knowledge and technology to mitigate or further adapt to these changes.

Because of its close involvement in the physical, chemical, and biological processes of the Earth System and its strong interaction with human social and economic activities, water plays a major role in the ongoing process of change. Water is inextricably linked to the future of humanity.

1.2   Development, Environment, and Water

Following World War II, changes in the world have accelerated in a number of ways. On the positive side, the growth of science and technology and the increase in the production and supply of goods and services have been remarkable. Political and social conditions have, generally, changed for the better. A significant improvement in the standards of living, in physical and other dimensions, for hundreds of millions of people has been a tangible result.

Thus, there has been an explosion of human enterprise, with important associated global-scale consequences on the Earth System. This unprecedented event was described by Steffen et al. (2007) as the "Great Acceleration" (see Box 1.1). The driving forces of the Great Acceleration form an interlinked system consisting of population increase, rising consumption, availability of abundant cheap energy, and liberalizing political economies.

In the core of the Great Acceleration, the 1960s, the Golden Sixties, brought the belief that science could and would solve all problems on Earth. However, in the 1970s some clouds appeared—the oil crisis of 1973, pollution of land, rivers, and seas, the arms race… The 1992 Rio Summit gave us ecological awareness on a global scale.

The end of the Cold War in the early 1990s brought back signals of hope—the democratic system had won and everything should be easier. However, the recurrent escalation of oil prices has raised awareness that the era of cheap oil could be over. In fact, it was the low cost of oil that fueled the development of western societies for almost two centuries, since the beginning of the industrial revolution.

The Brundtland Report, published in the late 1980s by the World Commission on Environment and Development (WCED 1987), introduced the concept of Sustainable Development, which became a popular slogan among academics, politicians, and the common citizen. However, implementing this concept has been a challenge, in part because of the difficulties of defining and identifying appropriate indicators for it. It appears that the end of an era of prosperity for western countries is currently being foreshadowed by the depletion of readily accessible and inexpensive oil reserves, the occurrence of climate change, overwhelming pollution, the dramatic

Box 1.1 The Great Acceleration

The human enterprise suddenly accelerated after the end of the Second World War. Population doubled in just 50 years, to over six billion by the end of the twentieth century, but the global economy increased by more than 15-fold. Petroleum consumption has grown by a factor of 3.5 since 1960, and the number of motor vehicles increased dramatically from about 40 million at the end of the War to nearly 700 million by 1996. From 1950 to 2000 the percentage of the world's population living in urban areas grew from 30 to 50 % and continues to grow strongly. The interconnectedness of cultures is increasing rapidly with the explosion in electronic communication, international travel, and the globalization of economies.

The pressure on the global environment from this burgeoning human enterprise is intensifying sharply. Over the past 50 years, humans have changed the world's ecosystems more rapidly and extensively than in any other comparable period in human history. The Earth is in its sixth great extinction event, with rates of species loss growing rapidly for both terrestrial and marine ecosystems. The atmospheric concentrations of several important greenhouse gases have increased substantially, and the Earth is warming rapidly. More nitrogen is now converted from the atmosphere into reactive forms by fertilizer production and fossil fuel combustion than by all of the natural processes in terrestrial ecosystems put together…

The lessons absorbed about the disasters of world wars and depression inspired a new regime of international institutions that helped create conditions for resumed economic growth. The United States in particular championed more open trade and capital flows, reintegrating much of the world economy and helping growth rates reach their highest ever levels in the period from 1950 to 1973. At the same time, the pace of technological change surged. Out of World War II came a number of new technologies—many of which represented new applications for fossil fuels—and a commitment to subsidized research and development, often in the form of alliances among government, industry, and universities. This proved enormously effective and, in a climate of renewed prosperity, ensured unprecedented funding for science and technology, unprecedented recruitment into these fields, and unprecedented advances as well.

The Great Acceleration took place in an intellectual, cultural, political, and legal context in which the growing impacts upon the Earth System counted for very little in the calculations and decisions made in the world's ministries, boardrooms, laboratories, farmhouses, village huts, and, for that matter, bedrooms. This context was not new, but it too was a necessary condition for the Great Acceleration.

The exponential character of the Great Acceleration is obvious from our quantification of the human imprint on the Earth System, using atmospheric CO_2 concentration as the indicator. Although by the Second World War the

(continued)

Box 1.1 (continued)

CO_2 concentration had clearly risen above the upper limit of the Holocene, its growth rate hit a takeoff point around 1950. Nearly three-quarters of the anthropogenically driven rise in CO_2 concentration has occurred since 1950 (from about 310–380 ppm), and about half of the total rise (48 ppm) has occurred in just the last 30 years.

Source: Steffen et al. (2007).

decrease in biodiversity, and the serious depletion of natural resources. Water is of special concern since it is linked to the life of humans and ecosystems, as well as to social and economic development.

The financial and economic crises that began in 2008, and the difficulties in coping with it, have led some to argue that environmental concerns lie at the heart of more sustainable social and economic development (Sachs 2008), and that this will require new development paradigms, involving radical changes in human behavior and in the evolution of standards of living.

Unfortunately, many still believe that, if the economy is stabilized, everything will return to business-as-usual, creating the right conditions to handle social and environmental problems. Others, however, think that there will be no choice, but to carry out profound reforms that might guarantee an effective global governance of the financial and economic systems. And they feel that it is important not to sideline the necessary social and environmental global reforms when designing and implementing economic policies.

From the onset of the industrial revolution two centuries ago, industrialization has brought about a new model of civilization. Its main drivers—technological development, agriculture, urbanization, and transportation—responding to population growth and increasing *per capita* demands, have promoted levels of consumption of water intensive goods and services that now seem unsustainable. Climate change is further exacerbating this situation. Many natural resources are, at present, being depleted or degraded faster than they can be renewed. We are, in fact, using up our natural capital, placing at risk our future prosperity and our very survival. We are losing control of the feedback mechanisms that may result from our actions. Climate change is a good example. In just two centuries, we have transferred to the atmosphere, in the form of gases and heat, a very substantial part of the hydrocarbons that have taken millions of years to accumulate on Earth.

Toward the end of the twentieth century, there were signs that the Great Acceleration could not continue in its present form without increasing the risk of crossing major thresholds and triggering abrupt changes worldwide. Transitions to new energy systems will be required. There is a growing disparity between the wealthy and the poor, and, through modern communication, a growing awareness by the poor of this gap. Globally, there are heightened material aspirations—a potentially explosive situation. The climate may be more sensitive to increases in carbon dioxide and may have more inertia than earlier thought, hinting at abrupt and irreversible changes in the planetary environment as a whole. This may adversely affect our natural resources and, in particular, our water resources.

1.3   The Anthropocene and Water

In 2000, the American biologist, Eugene Stoermer, and the Dutch geochemist and Nobel laureate, Paul Crutzen, considering the evidence and extent of recent human activities with significant global impact on the Earth System, concluded that it is more than appropriate to emphasize the central role of humankind in geology and ecology. They proposed to use the term "Anthropocene" for the current geological epoch (Crutzen and Stoermer 2000; Crutzen 2002).

Crutzen (2002) states that for the past two centuries, the effects of humans on the global environment have escalated. Because of the anthropogenic emissions of CO_2, global climate may depart significantly from natural behavior for many millennia to come. It seems appropriate to assign the term Anthropocene to the present, in many ways a human dominated, geological epoch, supplementing the Holocene—the epoch corresponding to the last 11,700 years, when climate has remained remarkably stable.

The Anthropocene is a dynamic state of the Earth System, characterized by global environmental changes, already significant enough to distinguish it from the Holocene, but with a momentum that continues to move it away from the Holocene at a geologically rapid rate (Steffen et al. 2011).

Fossils, or artifacts that are used to distinguish between different geological periods, may now be associated with human endeavor. Even cities will be particularly distinctive fossils (*The Economist* 2011). There is, thus, a strong dimension of urban dynamics, and what is referred to as "urban culture," which generates physical and other artifacts in situ as well as on a wider global scale, associated with the contemporary development trajectories.

The decision to create a new geological epoch requires an international geology congress to issue an official decision on the matter. In the meantime, an Anthropocene Working Group has been established, under the International Commission on Stratigraphy, to examine and debate the case for formalizing the term Anthropocene within the Geological Time Scale (Zalasiewicz et al. 2011). It might take years, or even decades, for the International Union of Geological Sciences, the world's geological governing body, to formalize the new epoch.

Even without such a formal declaration, it has become clear that we are now living in a world where humans are having a definite and global influence on major climatological, geophysical, and hydrological dynamics. The reaction of climate to greenhouse gases and the magnitude of the associated biosphere changes lend credibility to the idea that we have indeed entered into a new epoch that is not comparable to any interglacial episode of the Quaternary. The combination of extinctions, world species migration, and the general substitution of natural vegetation by crops grown under monoculture induces persistent bio-stratigraphic signals. Thus, evolution itself will be directed along new pathways.

In simple terms, it has become clear that humans have not only become big predators and big resource squanderers but also a global geophysical force, similar to some of the great forces of Nature, such as volcanoes or earthquakes (Williams et al. 2011). Population growth is only part of the issue. Humans, as a species, are also becoming wealthier and consuming exponentially more resources. In conclusion, we have become the main force on Earth, altering its natural balances. As Steffen et al. (2011, 757) state "We are the first generation with the knowledge of how our activities influence the Earth System, and thus the first generation with the power and the responsibility to change our relationship with the planet."

Crutzen (2002) noted that the Anthropocene could be said to have started in the latter part of the eighteenth century, when analyses of air trapped in polar ice showed the beginning of global concentrations of carbon dioxide and methane.

Our planet no longer functions in the way it once did. The Earth in which we live is currently confronted with a new situation, an unexpected danger—the proliferation of an endemic and invasive species (the human species) whose influence has transformed the atmosphere, impoverished the biosphere, changed the lithosphere, and largely modified the hydrosphere.

In fact, water is becoming a central issue in the Anthropocene. The currently growing concern covers not only fresh and transition waters but also the oceans, changes in which, in turn, also influence inland waters. In particular, climate change impacts affect water quantity (availability and demand) and quality, and also cause the sea level to rise, with non-negligible consequences to inland surface and ground waters, as well as on estuarine and coastal waters. Moreover, climate change influences rainfall patterns and tends to magnify extreme phenomena, such as floods and droughts (IPCC 2007, 2012).

Freshwater is a finite and mostly renewable resource. It is increasingly threatened by human action. Not only do we contaminate and pollute our rivers, lakes, and aquifers, but we also withdraw an increasing amount of water from them to meet the demands, in part because of population growth and the continued rise in standards of living. In addition, climate change and the development of potential flood zones have increased the incidence and impact of water-related disasters. According to the World Health Organization, during the last decade of the last century about two billion people were victims of natural disasters, of which 85 % were floods and droughts.

The challenges of the twenty-first century—resource constraints, financial instability, inequalities within and between countries, environmental degradation—also signal that business-as-usual cannot continue. We are moving into a new phase of human experience and entering into a new world that will be qualitatively and quantitatively different from the one we have known. The emerging Anthropocene world is warmer with a diminished ice cover—more sea and less land, changed rainfall patterns, modified river flows, disturbed groundwaters, and impoverished aquatic ecosystems, biosphere, and human-dominated landscapes.

Effective planetary stewardship must be achieved quickly, as the momentum of the Anthropocene threatens to tip the complex Earth System out of the cyclic glacial-interglacial environment in which *homo sapiens* has evolved and developed. And the evolution of water-related issues will strongly influence the new planetary stewardship.

1.4   The Increasing Water Crisis

For several decades, especially since the second half of the twentieth century, the rate of increase in water use on a global scale has been more than twice that of population growth (see Figs. 1.1 and 1.2). The uncertain nature of these trends and the predictable growth of population and gross domestic product (GDP) seem to lead to more, and larger, regions in the world being subject to water stress (see Chap. 2) and, if current trends continue unabated, to the development of increasingly unsustainable water use situations.

Current water demand and supply pressures are changing. Demand pressures include population growth and an increase in water intensive diets as a portion of the population moves into increasingly higher water consumption behaviors. Demand pressures also include growing urban, domestic, and industrial water use. Climate

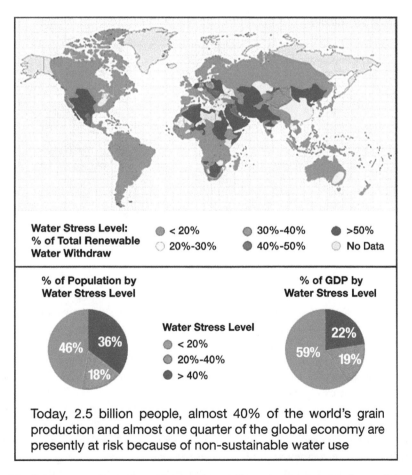

Fig. 1.1 Current water stress given present water supplies and management and use policies and practices. *Source*: Illustration courtesy of Growing Blue: www.growingblue.com

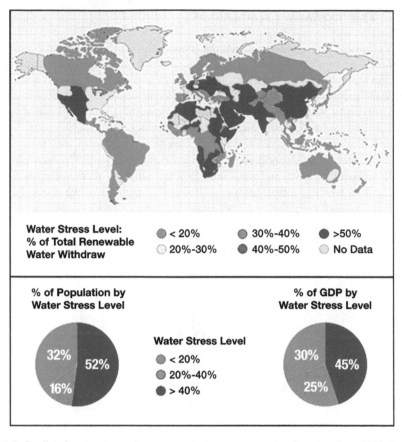

Fig. 1.2 Predicted water stress given current water management and use trends to 2050. *Source:* Illustration courtesy of Growing Blue: www.growingblue.com

change plays a role by creating additional water demand for agriculture, reservoir replenishment, and downstream flow augmentation for quality improvement, including temperature control. On the supply side, issues such as water transport, availability, and variability present challenges, as do the declines in renewable water supplies and nonrenewable groundwater resources.

In nearly every one of these categories, trends are moving in the exact opposite direction necessary to sustain future growth. Taken together, these trends create today's "water stresses" as shown in Fig. 1.1. And the resulting ecosystem pressures, along with current economic and political conflicts, only exacerbate these stresses.

Today, many regions of the world are already water stressed because of population and economic growth. In fact, 2.5 billion people (36 % of the world's population) live in these regions and 22 % of global GDP is already being produced in risky, water-scarce areas, affecting production as well as corporate reputations when competition over water use develops.

The amount of water directly used *per capita* varies considerably from country to country. In developed countries one can assume an average value of 200 L per person per day, with much higher values in the United States. The value adopted internationally for the amount of water necessary to cope with basic water needs is about 50 L per person per day (Gleick 1996). The amount of water used indirectly is much higher, from less than 3,000 to about 5,000 L per person per day, depending on a number of factors, in particular the diet system. This is the volume of water involved in the production of goods, which is especially relevant in the case of agricultural goods.

By 2050, the world will have to feed an additional 2–2.5 billion people. In order to sufficiently meet the nutritional needs of this additional population, we should consider the amount of water that is consumptively used in the production of different goods and, in particular, food.

There are four main factors that are particularly critical in the Earth's system: water, energy, food security, and climate change. These are all connected through interactions and feedbacks. For example, the growing, transportation, processing, and trading of food products require large amounts of water and energy. Burning fossil fuels or slash burning forests for agriculture can alter the climate system. Food security, agriculture, and water are also inextricably related. A complete analysis is provided by the Comprehensive Assessment of Water Management in Agriculture (IWMI 2007). This work demonstrates that in a business-as-usual scenario, water consumption in agriculture would increase almost twofold from approximately 7,000 to 13,000 km^3. Where this extra water will come from is highly uncertain. The above-mentioned interactions and feedbacks profoundly affect the hydrological cycle, which is further influenced by climatic variability, creating difficulties to all forms of life on the planet.

The increase in the demand for water and food may cause political and social instability, geopolitical conflicts, and irreversible environmental damage. Any strategy focused only on part of the water-food-energy system, without considering the strong interactions at play, risks having unexpected and serious consequences. The consideration of the water-food-energy interdependencies ensures that sector-based decisions take into consideration the intersector perspectives. This makes possible trade-offs among sector interests and favors synergies.

Water is increasingly coming to the forefront as a priority in our future. The third United Nations World Water Development Report (UN WWAP 2009) warns, in an unprecedented fashion, that extremely serious consequences may result from the current inequitable, unsustainable use of water. Both economic development and security are placed at risk by poor water management. That is why the well-established concern about a global energy crisis has recently begun to be accompanied by a concern about the well-documented, but often ignored, looming global water crisis. The energy and water nexus, expressed both by the effects of water use on energy consumption and by the effects of energy production on water consumption, is gaining increasing attention. In addition to the constraints on water referred to above, it must also be noted that the current development of biofuels will result in greater water consumption (and greater farmland use), which competes with water for food production. In fact, there is an important water, food, energy nexus (see, for example, Hoff 2011; WEFWI 2011).

Demographic growth, economic development, urbanization, and climate change have increased the pressure placed by humans on the water supply. However, it should be noted that the situation in developed and developing countries is considerably different. In many developed countries, industrial development and the awareness of water issues have implied that the economic activity of these countries is currently less conditioned by water than some decades ago, as a consequence of a more efficient water use. Developing countries are still at a phase where their water use is increasing in line with their growing population and/or economic development. This is further exacerbated by poor land use planning, the impacts of climate change, and the increased use of water in agriculture and animal husbandry related to the production of food and to the creation of jobs.

There are still huge gains to be made as the result of increased efficiency in water use. But there are limits. The efficiency in domestic and industrial water use can be improved through current and foreseeable technological development. There is also room for significant improvements in the productivity of water use in the agricultural sector. However, substantial amounts of water are consumed through evapotranspiration in agriculture. This is particularly important because water for agriculture corresponds, on average, to 70 % of total global water allocations and to over 80 % in many low and middle income nations. It is also clear that climate change and population increase and/or economic development will only worsen this situation. These increased pressures could perhaps be managed through adequate measures to mitigate climate change and enable adaptation with the help of science and technology, but such measures are not being widely pursued at present. Considering that the improvement in water productivity that has been achieved in recent years in some agricultural regions could be extended globally, and accepting that substantial improvements can be made in the efficiency of the supply chain, then available water resources should, perhaps, be able to cope with the 35 % increase in the world's population that is expected over the next 4 decades.

However, much higher pressures are expected to develop because large segments of the populations in the emerging countries will tend to raise their standards of living considerably.

1.5   Sustained Prosperity While Controlling Water Use

Prosperity is understood to be a successful, flourishing, or thriving condition, a state in which things are perceived to be going well. Prosperity often encompasses wealth and also includes other factors which are independent of wealth to varying degrees, such as happiness and health (Jackson 2011).

Prosperity is frequently measured through the increase of GDP *per capita*. This can be approximately true for the poorest segments of the world's population. However, it does not seem to be the case for the more affluent societies. As Jackson (2011) states, beyond a certain point, continued pursuit of economic growth does not appear to advance, and may even impede happiness. Our society has to be able to create the conditions under which it is possible to increase prosperity, within the ecological limits of a finite world.

Actually, a satisfactory conceptualization of prosperity has to consider the issue of limits. In fact, a continual economic growth of the type we have been adopting in recent decades is not compatible with the limited amount of the resources existent in the planet, water included.

Gleick and Palaniappan (2010) defined "peak water" and warned of situations where, similarly to peak oil, we can face growing threats of water scarcity, seriously restraining water consumption. An example is the unsustainable use of fossil groundwater, where we may consider peak water as entirely similar to peak oil. Another is the concept of peak "renewable" water, where the total renewable flow of a natural system is used by humans, and thus imposes limits on future water availability even though we are considering a renewable resource. In fact, there are definite limits on the water provided by the hydrological cycle in the world. There will be increasing difficulties in coping with the increased water demand, driven by demographic pressures and the increased *per capita* water demand.

Humans have difficulty dealing with future water problems, particularly at a global level. This is illustrated by the slow evolution of the positions adopted by international organizations at an international level. A mere 20 years ago, in the final declaration of the UN Conference on Environment and Development held in Rio de Janeiro in 1992 (UN 1992) water was barely considered. Water has only started to be considered as a crucial element of sustainable development since 1998, when the UN Commission on Sustainable Development adopted the text "Strategic Approaches to Freshwater Management" (UN 1998, p.2).

In the future, problems related to water at a global, regional, and local scale will appear increasingly complex and the conditions of access to water increasingly unequal. Water has now become a key issue for society and the newly emerging problems must inevitably act as catalysts for the search for innovative social, technological, and economic solutions.

Complex water systems are, in a certain measure, adaptable to changing human pressures, but, in some cases, they are also vulnerable. Aquatic ecosystems may react slowly to changes in water systems, or reach tipping points above a certain level of pressure, changing rapidly from an apparently stable state to an unstable one. The consequences of this kind of situation may prove difficult to recover from, or may even be irreversible. In the current context of change, the fragile defenses of ecosystems tend to be weakened, because of factors such as agricultural development, industrialization, urbanization, and climate change, thus making the management of water and ecosystems increasingly difficult.

The availability of water of adequate quality will tend to decrease because of the constraints imposed by the above-mentioned drivers. And this can limit our well-being in the future and our capacity to improve it, i.e., our prosperity. Water may play an important role in this limitation of prosperity. The fundamental issue is to attempt to reconcile our desires for a good life with the constraints imposed by the availability of a renewable, but limited, resource.

The Malthusian analysis, published over 200 years ago in the famous *Essay on the Principles of Population*, has been largely criticized and considered to be wrong. However, Malthus may have only been wrong about the timing of his previsions, because he did not correctly foresee the strong development of science and technology that, during a certain time, was successful in delaying the process. But he was insightful as regards the prediction of the type of situation which we are facing today, with the human population being increasingly pressed against the finite limits of the planet.

The question of ecological limits to growth was explored seriously for the first time in the early 1970s by a group of scientists commissioned by the Club of Rome (Meadows et al. 1972). The report analyzed the issue of the limits imposed on growth by the scarcity and degradation of natural resources. And their predictions turned out to be considerably accurate, in spite of the fact that when the report of the Club of Rome was produced, the existing database on natural resources and their depletion was still very limited. The report forecast significant resource scarcities for the early decades of the third millennium if action is not taken to restrain material consumption.

1.6   Important Future Water Issues

Given today's accelerated pace of human development and the slow pace of managing issues as complex as water resources, tomorrow's challenges are already at our door. Whether improving our governance policies or our infrastructure systems, years and even decades (not weeks or months) are required to implement change.

This time lag is especially troubling given that 52 % the world's population—and approximately half the global grain production—will be at risk because of water stress by 2050, if the status quo, "business-as-usual," behavior is followed. In addition, as shown in Fig. 1.2, 45 % of total GDP will be at risk because of water stress by 2050. That is one and a half times the size of today's entire global economy (IFPRI 2012).

It is foreseen that, within a couple decades, water scarcity may affect about two-thirds of the world's population. In many countries, there is still a tendency to deal with water scarcity problems, mainly by augmenting the *water supply*, e.g., by increasing surface and groundwater storage and allocation through the creation of new infrastructure, desalination of saltwater or brackish water, reuse of wastewater, or recharging aquifers. This tendency has prevailed over reducing *water demand*, e.g., by stemming the losses in transport and distribution systems, implementing adequate tariff systems, which seek to encourage lower water demand levels, changing water use technologies, and, generally, increasing the efficiency of water use in domestic, industrial, and irrigation systems; in other words, seeking to increase overall water productivity.

Reducing water demand can also be achieved by controlling other aspects that are not directly related to water, but which are equally important. This can be achieved, for example, by controlling demographic growth, increasing the efficiency of the use of goods that consume water (in particular food products) in their production processes and along their supply chains, promoting appropriate land use planning, or attenuating the effects of climate change on water through adequate mitigation and adaptation measures.

The growing complexity of water management implies that the water managers of the future will have to be familiar with a wide range of disciplines and be able to interact with a variety of professionals as well as users. These managers and water management institutions should have sufficient technical, economic, social, financial, and environmental skills to be able to engage in dialogue with the professionals of these areas who are part of the teams responsible for water management. Furthermore, they should have the capacity to interact with politicians in order to understand their short-term political commitments, as well as to facilitate the conciliation of politicians' initiatives with long-term sustainable water resource policies.

Within the context of the changes referred to above, it is crucial that the different countries and groups of countries sharing a certain river basin review their water management strategies, in order to meet water demand in a sustainable manner and ensure a balance between water supply and demand, as well as achieving a balanced use of surface water and groundwater. The very important case of transboundary river basins must be an object of a special consideration, as water is, in large measure, a transboundary resource, which requires for its adequate management the existence of institutions that cross traditional political borders.

An important future challenge will concern increasing our capacity to manage water and aquatic ecosystems, taking into consideration the resilience of social and ecological systems within the framework of the sustainable development models to be adopted in the future.

1.7   Water and Globalization

In recent years, globalization has promoted new rules and procedures for the international trade of goods and services, reflecting an increasing influence of multinational firms. This globalization of trade has wide-ranging implications for consumers, governments, and the environment.

The relevance of globalization as regards water may be considered from two different perspectives. The first is related to the consequences of economic globalization of water resources management. The growing integration of the world economy is the source of dynamics that tend to extend beyond traditional borders and can have negative effects on water, in particular concerning water contamination and associated environmental degradation.

The second perspective is related to water itself as an object of global trade policies. Some natural resources, such as oil, natural gas, wood, agricultural products, or fish have, for a long time, been traded in international markets without becoming a political issue. When it comes to the export and import of water, however, more expressive reactions tend to develop. Water is, in fact, different than many other natural resources that are traded because the costs of transport are very significant in comparison to the economic value of water and because of perceptions about the human right to water, and objections to the commodification of the resource. Gleick et al. (2002) and Hoekstra and Chapagain (2008), among others, have dealt extensively with the issue of water and globalization.

The international trade of certain resources, such as agricultural products, livestock, fossil fuels, fish, and lumber, always involves a certain degree of processing. Other resources, such as crude oil, wood logs, or raw fish, involve a much smaller degree of processing at the point of origin. While most water is not traded internationally because of the high cost of transportation, water can, in some cases, be produced as a product with high added value. Bottled water, for instance, is a product with significant added value and its international trade has been growing considerably in numerous regions.

International projects involving water transfer often raise concern and controversy. However, one form of "trade," which is generally accepted without raising special problems, is the natural flow of water between countries sharing a river basin. This transaction is normally ruled by political agreements, rather than trade agreements.

In practice, only a comparatively small number of agreements for the long distance trade of raw water have been concluded. The international trade of raw water is normally made through pipes or canals or by transporting water in tankers or large plastic bags—called "medusa bags"—towed by a boat (Gleick 2000b). The transport of water by towing icebergs from the polar regions wrapped in plastic bags has also been proposed but has never been put into practice. These forms of water transfer are very expensive and usually pursued only in rare cases where other practices, such as desalination, are not possible. Almost all such efforts only provide water for very high-value industrial or domestic needs and not for other important uses, such as food production.

In the future, global, regional, and local water issues will tend to become increasingly complex and the conditions of access to water progressively unequal. Water will be a key issue for society and new problems will encourage the search for innovative solutions requiring expertise in the social, technological, and governance areas.

1.8   Water Governance

The implementation of an economically efficient, socially equitable, and environmentally sustainable water resource management policy can only be achieved through good governance. Recently, the concept of effective water governance has grown in importance and has led to the widening of the water agenda, so as to include the consideration of democratization processes, corruption, and power imbalances between poor and rich countries and between rich and poor peoples.

Historically, the current concept of water governance appeared for the first time in 2000, at the Second World Water Forum in The Hague in 2000 (WWC 2000). However, the Forum Declaration referred to *good water governance* as water resource management involving public interest and stakeholder participation, which is a comparatively narrow definition. Only one year later, at the Bonn Freshwater Conference (GIZ 2001), the concept of water governance included institutional reform, legal framework, equitable access, and integrated water resources management. Since then, the term "water governance" has definitely become a part of the vocabulary of water professionals.

According to the definition put forward by the Global Water Partnership in 2003, *water governance* refers to "the range of political, social, economic and administrative systems that are in place to develop and manage water resources, and the delivery of water services, at different levels of society." (Rogers and Hall 2003, 16). Box 1.2 presents an overview of water governance concepts, as provided by OECD (2011).

Water governance determines who gets water, from where, and when and how; it also establishes who has the right to water and to water services. Water crises would thus be the result of inadequate governance, more than actual water scarcity. As Rogers and Hall (2003, 7) state, "governance relates to a broad social system of governing, which includes, but is not restricted to, the narrower perspective of *government* as the main decision-making political entity." Managing water, as part of the socioeconomic system, is conditioned by general politics and is, thus, influenced by decisions that lie outside of the water management agencies. The representation of various interests in water decision-making and the role of politics are important in the definition of governance dynamics (UN WWAP 2009).

Only recently has global water governance started to attract the interest not only of the natural scientists and engineers but of social scientists as well. Achieving effective water governance involves a wide range of issues that have been the subject of many authors' work. One of the most important is the implementation of *integrated water resources management*. This has been defined as "a process which promotes the coordinated development and management of water, land and related resources, in order to maximize the resultant economic and social welfare in an equitable manner, without compromising the sustainability of vital ecosystems" (GWP 2000, p.22). In the past, water management was concentrated on sectors and there has been little coordination. Integrated management should focus on the interactions between the natural system (which has an effect on the availability of water and its quality) and the human system (which affects water use, pollution, wastewater production, and also the definition of development priorities). However, OECD (2011) recognizes that integrated water resources management as adopted by many countries cannot be properly implemented without considering a broader governance framework. This would include not only sustainable water policies but also measures governing scientific, educational, and technological issues, as well as communication and participation.

Water governance is considered to have four different dimensions—social, economic, environmental, and political—which have been identified by UN WWAP (2006) in the following way:

- *The social dimension* points out the equitable use of water resources. Apart from being unevenly distributed in time and space, water is also unevenly distributed among the various socioeconomic strata of society, in both rural and urban areas. How water quantity and quality and related services are allocated and distributed directly affects people's health, as well as their livelihood opportunities.

Box 1.2 Overview of Water Governance Concepts

The term *governance* is now used widely by governments, international organizations, private operators, civil society, donors, and aid agencies, but it has been defined in different ways. It originally served to connect debates on politics and administration that equated governance with government, but the focus has subsequently been extended beyond government to encompass relationships between a range of state and non-state institutions (Kaufmann et al. 2006). *Public governance* now refers broadly to power and authority and to how a country manages its affairs. It is taken to encompass all the mechanisms, processes, and relationships that institutions, citizens, and groups use to articulate their interests and to exercise their rights and obligations.

The terms are sometimes used interchangeably, but *water governance* should be distinguished from *water management*. Water governance formally refers to a set of administrative systems, with a core focus on formal institutions (laws and official policies) and informal institutions (power relations and practices) as well as organizational structures and their efficiency. Water management covers the operational activities for meeting specific targets, such as aligning water resources and water supply, consumption, and recycling. Institutional and policy frameworks that foster transparency, accountability, and coordination are, thus, part of good water *governance*. Delivering water or installing improved water services are part of water *management*. Despite the extensive literature on the topic, the consensus associates water governance with "doing things right."

This encompasses several dimensions, such as economic governance, corporate governance, international governance, regional governance, national governance, and local governance (Dixit 2009).

Most governance principles for managing water resources and services are based on common pillars. They have been variously combined in different frameworks, thus emphasizing certain universal aspects of governance (Lockwood et al. 2008):

- *Legitimacy* of the organization's authority to govern
- *Transparency* in the decision-making process
- *Accountability* of actors and their responsibilities, including integrity concerns
- *Inclusiveness* of the different stakeholders
- *Fairness* in the service delivery or allocation of uses
- *Integration* of water policy making at horizontal and vertical levels
- *Capacity* of organizations and individuals managing water
- *Adaptability* to a changing environment
  Source: OECD (2011).

- *The economic dimension* draws attention to the efficient use of water resources and to the role of water in overall economic growth. Prospects for aggressive poverty reduction and economic growth remain highly dependent on water and other natural resources. Studies have illustrated that *per capita* incomes and the quality of governance are strongly and positively correlated across countries.
- *The environmental dimension* shows that improved governance allows for the enhanced sustainable use of water resources and ecosystem integrity. The appropriate regimes of water of good quality and sediment are critical to maintain ecosystem functions and services and to sustain groundwater aquifers, wetlands, and other wild-life habitats.
- *The political dimension* points at granting water stakeholders and citizens-at-large equal democratic opportunities to influence and monitor political processes and outcomes. At both national and international levels, marginalized citizens, such as indigenous people, women, and slum dwellers, are legitimate stakeholders in water-related decision-making.

Obviously the role of stakeholders is important as they influence the above-mentioned dimensions of water governance. Better water governance is now recognized worldwide as a key condition for a fairer, cleaner, and greener economy. In the context of the global water agenda, improvements in water policy are likely to be very limited, or at best slow and incremental, unless water institutions become more effective both independently and by interacting with each other. Innovative policy is called for, but institutional responses must also be designed, especially in the current context of financial constraints and climate change. Governments in countries where water is plentiful, as well as those where water is scarce, must now do better with less and less resources (OECD 2011).

1.9   Toward a Global Water Governance?

Water governance refers to different geographical levels of aggregation:
- *The local level*, taking into consideration local needs and stakeholders' views
- *The river basin level*, emphasizing the importance of natural water boundaries for efficient water governance
- *The regional level*, encompassing several river basins in a given region to take into consideration national perspectives and/or land use planning
- *The global level*, that points out the importance of considering long distance effects on water availability and consumption

The relationships between these four levels of governance and a definition of how these relations can be used to strengthen the adaptive capacity and the resilience of water systems deserve some attention.

The river basin level is generally accepted as the natural reference framework for the consideration and resolution of water problems, through the actions of planning and management, which are considered as part of water governance. The local level is not normally sufficient, as upstream-downstream relations condition most of the water issues and involve both water quantity and quality aspects.

The relationship of water governance with the governance of the various economic sectors of society, in harmony with a general governance of society, may imply consideration of a level of water governance above that of the river basin. In other words, in certain cases, water governance at the stream basin level may not be sufficient, making it necessary, perhaps, to try to reach a higher level—i.e., regional or even global. Much depends on the specific problems being addressed and the extent of these problems.

Global water governance may be needed if the hydrological cycle and its interactions with other biogeochemical cycles are being considered, or if many socioeconomic consequences of regional water use are felt globally. Such effects include those related to climate change, to the global distribution of food and other products, to global security, or if the water problems being considered are projected beyond the boundaries of regional water governance.

The production of goods in certain countries or regions of the world for consumption in different countries or regions implies that the water used for producing these goods is used up in parts of the world which are not the same as those areas where the products are consumed. In a way the water involved in the production processes can be considered as being transferred to the places where the final consumers of the products are located.

The amount of water used in the production of various foods is quite variable, depending on the product and on the conditions under which it is produced. As an example, the water associated with the production of 1 kg of wheat averages 1,300 L, whereas production of 1 kg of coffee uses an average of 21,000 L. If we consider the case of meat products, it can be mentioned, for instance, that producing 1 kg of chicken involves the use of 4,000 L, while 1 kg of grain-fed beef requires 15,000 L.

When making water balances in a country or a region, one generally speaks of the amount of water generated by rainfall in that country and the amount of

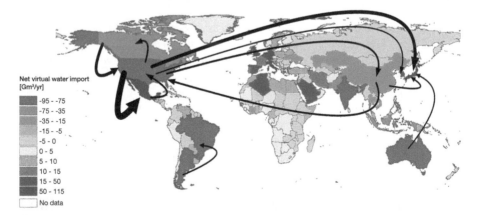

Net virtual water import [Gm³/yr]

- -95 - -75
- -75 - -35
- -35 - -15
- -15 - -5
- -5 - 0
- 0 - 5
- 5 - 10
- 10 - 15
- 15 - 50
- 50 - 115
- No data

Fig. 1.3 Virtual water balance per country and direction of gross virtual water flows related to trade in agricultural and industrial products over the period 1996–2005. Only gross flows in excess of 15 billion m³/year are shown. *Source*: Hoekstra and Mekonnem (2012)

water, both surface and groundwater, that flows in from neighboring countries. But if this balance is to include the water used in the production of goods, then imports and exports of water involved in the production of goods must also be accounted for.

Figure 1.3 shows the flows of virtual water involved in the production of agricultural and industrial products, over the period 1996–2005. According to Hoekstra and Mekonnem (2012), the region with the largest exports of virtual water is North America, the largest importers being in Central America and South Eastern Asia.

1.10   Creating a Water Secure World

It is impossible to define just one optimal path to take to reach one optimal future, say in 2050. There are many paths, and many possible "desired" futures. But, for sure, water scarcity will be one aspect that the more than nine billion people (as expected by 2050) will face, together with a strong increase of *per capita* water demand. This will particularly apply in emerging countries, such as China and India, which together by 2050 will represent one-third of the world's population. Such countries are reaching economic and educational levels that fully legitimize the desires of their citizens to approach the standards of living and consumption of the developed countries.

The current public and political concern with global warming is considerably stronger than the concern about a global water crisis. These two crises are closely interrelated. Global warming will impact water supply and demand and affect water quantity and quality. At the same time, water is clearly the main mediator of the effects of climate change in the economy, society, and environment. Moreover, this water and climate change nexus is influenced by the changes affecting other sectors, in particular energy and food production.

Furthermore, with increasing water demand and with more irregular water availability, environmental water requirements will also increase in importance. The demand for water has often been considered to be at odds with the need for water to maintain the life of ecosystems. In addition, discharge of human waste, and in particular wastewater, into the environment has often continued without special concern for its effects on ecosystems. We will have to address how depriving ecosystems of water that is essential to life and poisoning them with waste will impinge on human life and development. To achieve this outcome will demand an interdisciplinary and intersector analysis of the processes and issues involved in future water scarcity, which should include the interactions and feedback effects of water, energy, food security, and the health of ecosystems and the external drivers.

A basic premise of the analysis to be made is that, in the future, there will be sufficient water to feed a growing and wealthier population, to sustain vital life support systems, and to produce and distribute enough other goods and services. To achieve this, it will be essential to widen the notions of "more crop per drop"— often used in the agricultural context in order to include multiple water-related benefits—"more jobs per drop," "better environment per drop," "improved nutrition per drop," etc. However, many of the current trends imply that we could be heading toward more drops per job, or per given state of the environment, or per given level of nutrition, unless we work together to do otherwise.

Catering for unmet and growing human needs and escalating wants is a mounting challenge. To consider this and, at the same time, sustain the functioning of water (fresh and brackish/marine) and terrestrial ecosystems is a major issue in any strategy for a desirable future. Certainly there are alternative pathways ahead and political commitments must be in harmony with social acceptance and compliance and, of course, show respect for life support systems.

The threat of a global water resources crisis is both possible and very frightening. To continue to follow a business-as-usual pathway will lead to a very difficult situation, corresponding to a horizon beyond which our normal forms of reasoning and our current models may not work at all.

In order to face the current and coming problems of water, some form of cultural revolution is needed, in terms of the ways through which such problems are identified, analyzed, and resolved. We also need to link water management and governance to a better governance of the drivers that are the cause of the pressures on water.

To prevent a severe water crisis we will need to show extraordinary creativity. We need to set in motion systems of governance and regulation that are capable of ensuring long-term sustainable development.

We will also need solutions to confront simultaneously financial, social, and environmental challenges. Our main concern must be to create, with the utmost urgency, the necessary conditions for human creativity to flourish in the domains of politics, science, culture, and ethics. Finally, local, regional, and world leaders need to put water issues at the top of their agendas before they are forced to by circumstances that they can no longer control. Clearly the most urgent task is to identify and implement creative long-term solutions to the water management problems that could threaten the future of humanity.

The issue of global sustainability, very much interlinked with water, is fundamentally about the ways people live and interact with the Earth System at the local, regional, and global scales. The remainder of this book explores some of the dimensions of the water governance challenges and why it is important to address these challenges, now.

Drivers of Water Demand, Course Changes, and Outcomes

2

There is enough water, food, and energy to meet everyone's needs. Yet for the bottom billion (Collier 2007) they are not being properly met. Will everyone's needs be met in the future? The answer to this question depends on what strategic steps we take today and on the day-to-day actions of consumers and the public at large. Learning from what has worked in the past and what has not, is important. Humans have the capacity to generate unprecedented wealth, but we are doing it in an environment of growing vulnerability, uncertainty, and stiffer resource use competition. Efficiency in the use of the limited water and other resources is necessary. Resource stewardship needs to be combined with strategies that ensure that all people have access to water and also the goods and services that water provides. In this regard the needs of the bottom one billion and the wants of the wealthier populations pose a double challenge.

2.1 Staggering Growth, Water Use, and Human Behavior

Imagine if we had asked people 100 years ago what they thought would happen during the coming century, in terms of economic growth, demography, water, and the environment. The term "environment" might have been explained with reference to the four elements identified by Aristotle—water, earth, fire, and air. Naturally, predictions would have varied because it is always hard to imagine what the future will bring. Yet it is important to try to understand how the future may evolve so we can be prepared and benefit from it—or try to change course.

During the past 100 years—the twentieth century—the aggregate global gross domestic product (GDP) multiplied 19 times (IMF 2000). With a population increase of less than a quarter of the growth in goods and services during the same period, there was an unprecedented opportunity to cater to everybody's needs, while, at the same time, providing opportunities to meet human wants. Some of the goods and services that are included in the calculation of GDP do not contribute to human welfare, but the staggering increase on the supply side reflects an impressive

Gulbenkian Think Tank on Water and the Future of Humanity, *Water and the Future of Humanity: Revisiting Water Security*, DOI 10.1007/978-3-319-01457-9_2, © Calouste Gulbenkian Foundation 2014

increase in demand and purchasing power for hundreds of millions of people. But the benefits have not been shared by everyone. Gaps between the rich and the poor have been, and remain, wide and are increasing (IMF 2000; Shah 2010).

Access to safe water is still a dream for about 900 million people and more than 2.5 billion lack access to safe sanitation. Food insecurity is a dire reality for close to a billion. Improvements were achieved after the beginning of the 1960s, but during the last 15 years or so, food insecurity, in terms of undernourishment, has increased in spite of a continuous increase in aggregate global *per capita* food production and supply (Lundqvist 2010; Fig. 2.2). Mr. Henry Kissinger's laudable pledge at the first World Food Conference (later called World Food Summits) in 1974, "In 10 years no child will go to bed hungry." has not become a reality.

Increased food insecurity during the last 15 years or so took place despite increases in production and supply being faster than the population increases during the same period. Production, and also supply—what is available in the market after losses and conversions—are at a level that is in excess of what is needed for food security. Today overeating is a more widespread phenomenon than undernourishment. It is estimated that 1.5 billion persons, aged 20 years and above, overeat (Beddington et al. 2012). Moreover, between one-third and one-half of the food produced on the farm is lost, wasted, or converted between "field to fork" (Lundqvist et al. 2008; Lundqvist 2010; Gustavsson et al. 2011). Increasing resource use efficiency— "more crop per drop"—is important, but if a large fraction of the produce is lost or not beneficially used the net result of gains in production are reduced.

These kinds of imperfections not only apply to the use of water and food. Similar observations can be made regarding other resources and commodities. With the strong linkage between food, water, energy, the environment, and human well-being (Hoff 2011) these kinds of imperfections affect all of us as well as the natural resources and Earth's environment. We need fresh thinking. In an era of growing water scarcity we need to identify how to use our technical capacity and sound and responsible human action to reduce these production and supply chain imbalances and imperfections. Reducing the losses and waste associated with the products we grow or produce from water is important. So far it has not been seriously tried.

2.2 Human Ambition and Capacity to Modify Our Planet

Promises to improve water and food security, among other things, for the poorest one billion are repeatedly given at international and national gatherings. At the same time, most of us who have enough want more. With higher incomes, many have the means to realize their wants. Within a generation, an additional two billion people will express their claims in terms of access to water, food, energy, and a range of goods and services.

The interdependencies between social ambitions, human well-being, and political promises, on the one hand, and natural resources and the environment on the other, are obvious and determine the kind of development that is realistic and stable. Yet there is a disconnect between the production and supply of goods and services and the ability of our Earth to sustain this production and provide this supply. The tremendous expansion in the production and supply of goods and services in the recent past has meant jobs, income, and, generally, possibilities for a better life. It has also meant a heavy exploitation of natural resources, the generation of a number of side effects, and, generally, a modification of the Earth System at a level and speed that is now overshadowing geological and natural forces (Crutzen 2002).

While the quality of life and material standard of living for a majority of us has improved considerably, the effects on water and other vital components of the Earth System have moved in the opposite direction. Many river basins in the world are labeled as "closed" or are on the verge of being closed or reaching peak limits (Seckler 1996; Gleick and Palaniappan 2010). An estimated 1.4 billion people live in closed basins (Smakhtin 2008). The downstream segments of a growing number of rivers are occasionally dry as a result of the heavy withdrawals of water from the stream, e.g., Colorado, Nile, Amu/Syr Darya. A closed basin naturally has more limited development options, even if groundwater is available. As the availability of surface water becomes more and more constrained, the use of economically available groundwater tends to increase.

There is no escape from the fact that the need and demand for finite and vulnerable water will continue to expand and so will competition for it. More uncertainty in availability, higher frequency of extreme weather events, and more rapid return flows of water to the atmosphere are also to be expected in the future. Given the changes in the hydrologic cycle as a result of land use and climate changes and the closed character of many basins, allocations to, and patterns of future water use, for various sectors in society will deviate from past trends. It is critical to better understand how these complex interactions may develop and the associated social, political, and environmental implications over the coming decades. Clearly, water issues will become even more important in the lives and activities of people. Any vision of the future needs to recognize that water is everybody's business (Cosgrove and Rijsberman 2000).

2.3   Visioning Our Future

Here we speculate on how the need and demand for water will unfold up to 2050. To do this we must first summarize the key drivers of past development and assess how they are likely to guide the future demand for, and management of, water. We can then identify which circumstances may change the course of the development trajectories. This process is illustrated in Fig. 2.1.

In many studies about possible futures, a range of drivers are considered. For the Scenarios Project of the UNESCO World Water Assessment Programme (UN WWAP 2012), for instance, in-depth research was carried out on ten drivers to examine possible future developments and also to look for inter-linkages between the drivers. Gallopin (2011) summarizes the work presented in several studies and talks about driving forces grouped into clusters. Ten clusters are identified and within each of them, on average, from 5 to 7 specific driving forces, trends, or

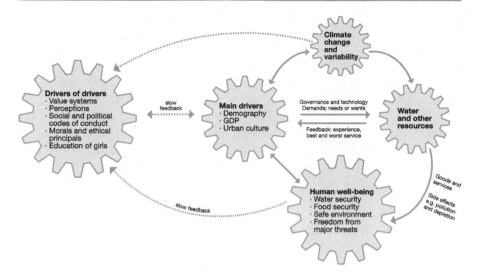

Fig. 2.1 An overview of the assumptions, contents, and purpose of this chapter (figure drawn by Britt-Louise Andersson, SIWI)

processes are identified. In other studies, a limited number of drivers are used. Kolbert (2011) used three variables—population, technology, and GDP—to estimate the dynamics behind the recent high rates of growth in many aspects of society, including GDP, energy, food, technology, and population (Steffen et al. 2011).

A common feature in these kinds of studies is that demography and GDP are regarded as the two main drivers of change. In addition to these two main drivers, a complex set of more basic, but also less concrete drivers, are at work. Social perceptions, political ideologies, and government strategies, knowledge, and value systems are examples of these kinds of underlying factors and forces. These are referred to as the drivers of the drivers in Gallopin (2011).

It is assumed that "Main drivers" and "Climate change" will increase in significance during the foreseeable future, whereas "Drivers of drivers," "Water and other resources," and "Human wellbeing" may either be more important as drivers or have a decreasing influence on the complex interactions in the system.

The main drivers are tangible and quantifiable whereas "drivers of drivers" are important, but hard to quantify. Demographic and economic trends can be projected with a relatively higher degree of likelihood as compared to changes in the underlying drivers. Many of these forces have typically been rather inert and their implications for water use and quality are difficult to evaluate. History provides, however, many examples of sudden and dramatic changes in, for instance, political systems. Contemporary political turbulence in North Africa and the "Arab Spring" illustrate that events in civil society may trigger processes of political change at national and regional scales. The distinction between main drivers and the indirect drivers is thus not necessarily reflecting which sets of drivers are more or less important.

2.3.1   Predicting the Future of the Planet

The effects of socioeconomic and political shifts on water demand and management are variable. Two examples shed light on drivers for policy change. The Water Act, introduced in South Africa in 1998, is often associated with the new political regime after 1994. However, by the early 1980s it had become evident that demographic and water use trends in South Africa had resulted in serious problems that could not be solved within the existing policy framework. Mackay (2003, p.51) explains how members of the scientific community realized early the need for policy reform, but "…they did not achieve high priority on the national agenda until 1994." In addition, a widespread social discontent with the old regime paved the way for a new national agenda, which included a radically new perspective on water management.

Another kind of a water challenge is illustrated by the dramatic consequences of recent prolonged periods of drought followed by floods in Australia. As discussed in Box 2.1, the actors in this context are farmers, spokespeople for environmental sustainability, and government representatives. Pittock and Connell (2010) argue that the devastating drought in Australia can be seen as a "demonstration of the planet's future." Among other things, this example illustrates that the notion of "extreme events" refers to the painful and difficult social and political processes of adaptation to water and climate realities that may last for many years.

Obviously, a combination of circumstances, which vary from one country to another and over time, drive and constrain the management of water. A combination of socioeconomic forces, well-informed and articulate members of the scientific community, technological capacity, and climate context also play prominent roles in most countries. The political and administrative system is significant, but perhaps mainly as a means of enabling, implementing, and enforcing management plans and policies. The frequent reference to "political will" is relevant, but, as illustrated in Box 2.1, the critical issue is often "political skill," that is, a visionary and strong leadership where opposing interests are balanced, where policy is informed by scientific understanding, and decisions are negotiated and socially accepted (Lundqvist and Falkenmark 2010).

Box 2.1   Policy and Management Responses to the Prolonged Droughts that hit Australia, Especially in the Murray-Darling Basin 2000–2009

Australia has always been subject to major climatic variation. The history of the development of the water resources of the Murray-Darling Basin demonstrates just how water scarcity can arise from a combination of biophysical and socioeconomic factors. It also indicates how climate change and variability may affect food production.

(continued)

Box 2.1 (continued)

In the Murray-Darling Basin, the drought also brought to the forefront of public debate the fact that past government policies had over-allocated the basin's water resources to the detriment of the environment. Historically water entitlements were granted to irrigators by state governments. By the last quarter of the twentieth century, entitlements exceeded water availability in some sub-basins, although this was managed by allocating water on an annual basis based on storage in the dams. Floodplain forests and riparian vegetation communities were under increasing stress, migrating birds' habitats were being reduced, fish stocks and biodiversity were threatened, the salinity of the Murray River was rising, and the terminal Coorong wetlands were becoming more saline than the adjacent ocean. Furthermore, the Murray stopped flowing into the ocean for most of this period.

As a consequence of these events, environmentalists lobbied the state and federal governments hard to try and recover water for the environment. This was opposed by the farmers that stood to lose water. The drought had, however, already triggered a number of responses to water scarcity, which were further encouraged by water reform policies being overseen by the National Water Commission. These included the separation of land and water rights and the development of a market for water trading. This enabled irrigators to sell and purchase water on either a temporary (annual allocation) or permanent (entitlement) basis.

In terms of the governance of the system, the ongoing drought and associated environmental consequences triggered a rethink on how the Murray-Darling Basin waters should be governed. The Murray-Darling Basin Commission (MDBC) had administered the waters of the Basin for many decades. The MDBC was comprised of state commissioners who oversaw a secretariat based in Canberra. Under conditions where relatively drastic action was required, the lack of independence of the commissioners, who essentially represented their states, did not facilitate the responses needed to deal with water scarcity. Consequently, the Federal Government dissolved the MDBC and established an authority that reported directly to the Federal Minister of Environment and Water Resources. The new Murray-Darling Basin Authority (MDBA) was charged with developing a Basin Plan that would deal with the critical issues of water allocation.

The initial draft of the plan recommended very significant cuts to irrigation water allocations and created a major furor among irrigators when released in 2010. The Chairperson and Chief executive of the MDBA both resigned and, late in 2011, a revised plan was released. Using 2009 as a baseline year, the environmentally sustainable level of extraction (10,873 gigaliter/year or 10.87 km^3/year) would be achieved by reducing consumptive use of water by 2,750 gigaliter/year (2.75 km^3/year) (MDBA 2011). Of this, an estimated

(continued)

Box 2.1 (continued)

1,068 gigaliter/year (1.07 km³/year) has already been recovered for the environment through buyback by the Federal Government and infrastructure improvement schemes. A further recovery of 214 gigaliter/year (0.21 km³/year) was announced recently, leaving 1,468 gigaliter/year (1.47 km³/year) to be secured. There are further caveats indicating that this will not come into force until 2019 and may be further revised based on new scientific evidence. Unfortunately, it has not placated irrigators, who stand to lose significant quantities of water. Nor has it placated environmentalists, who have argued in the press and via political lobbying that even this reduction of allocations will lead to further degradation and decline of the system (MDBA 2011).

2.3.2 Feedback Mechanisms

The interactive and dynamic character of water demands coincides with increasing uncertainty about the amounts of water that are available in rivers and other water bodies. Extreme weather events, during short or prolonged periods, increase the risk and cost of water development and use.

Increasing temperatures are speeding up the hydrological cycle, as evidenced by more intense rainfall patterns in some regions and more rapid return flows of water to the atmosphere. With higher temperatures, agricultural seasons are affected. Similarly, an increased frequency of extreme events increases the risks for the users and managers of systems for water regulation and supply, e.g., dam operators. Higher air temperatures increase vapor content, which adds to the severity of extreme events (see Chap. 3). Opportunities for reuse of water in agriculture are likely to be further curtailed, i.e., within the time span of a season. Uncertainty is bound to increase.

Societies are responding by implementing mitigation and adaptation measures. Some of these are slow, others are fast; some are going in one direction whereas others result in interactions (illustrated by arrows in Fig. 2.1).

It is mind-boggling that one of the most dramatic changes that has ever affected society—global warming—and the close link to an increasingly precarious water situation—has not resulted in more concrete efforts to cope with the threats. The result, in terms of climate and water changes, is likely to have a huge effect on agricultural systems, natural habitats, and economic systems, in addition to the hydrological cycle itself. Scientific arguments about the seriousness of the consequences of our increased greenhouse gas emissions expressed at international climate meetings are only marginally and slowly influencing political decisions and concrete action. Social pressure on the political system to design and implement effective adaptation measures is weak.

The scientific truth is inconvenient not only for politicians but also for existing economic systems and even parts of the public at large. Opinion surveys indicate a widespread worry in countries, e.g., the member states of the Organisation for Economic Co-operation and Development (OECD), about climate change and its likely effects. But our readiness to modify behavior that is detrimental to our environment is a big challenge in any society.

2.3.3   Altering the Trajectory of Development Efforts

We offer some thoughts here on the importance of identifying the seeds for new thinking and how alternative or supplementary water management approaches could be formulated to gain the necessary political and social acceptance. The likelihood of a revision of established practices must be weighed against a tendency in society to stick to "sanctioned" thinking and entrenched prevailing practices. Many observers have noted a logic in the fostering of social and political behavior, usually referred to as path dependency (e.g., North 1994). Past investments and education tend to perpetuate a way of thinking among people, causing them to formulate and execute policies even if they are inferior to known alternatives. Investments made in water infrastructure (such as dams), storage, and conveyance facilities represent huge stocks of physical capital. The already existing structures and associated institutions and knowledge are, of course, very important for many people. Nevertheless, it makes sense to ask if alternative land and water management strategies may generate desirable livelihoods with less environmental risk for an adequate number of people, given the growing water constraints.

Path dependence is related to policy formulation. Hirschman (1975) eloquently argued that some problems attain a "privileged" status whereas other problems are "neglected" in policy. Projects and ideas in the former category receive strong positive attention and are tackled with "more motivation than understanding." It is natural that the public, as well as policy makers, are fascinated by the grand schemes in ancient hydraulic civilizations and similar schemes in the contemporary era—the Hoover Dam, Aswan Dam/Lake Nasser, Three Gorges project, etc. They are all magnificent showpieces. Once the planning of these structures has started, strong forces will drive the process and follow-up steps will naturally be taken. Loans are granted, equipment is made available, logistical support is forthcoming, etc. There is certainly a momentum that pushes for a continuation along the same trajectory. In comparison, rainfed systems are less spectacular and political interest and budget allocations are generally less enthusiastic.

2.4   Lessons from the Past

Hydraulic works have played a pivotal role in the development of societies and civilizations from ancient times. Artifacts from the so-called hydraulic civilizations still attract attention and admiration in, for example, Egypt and other parts of the world. The ideas that generated these grand schemes in ancient times are reflected also in the more recent past. A new wave of construction occurred during the previous century.

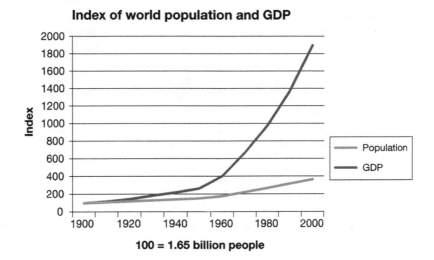

100 = 1.65 billion people

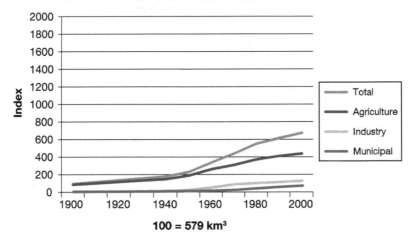

100 = 579 km³

Fig. 2.2 Growth of population and GDP (*top* graph) and water withdrawals (*bottom* graph) between 1900 and 2000. The situation in 1900 = 100 for all three variables. *Source*: Data taken from UN population estimates (UN 2004); various publications on water resources; IMF 2000; Madison 1995 (figure drawn by Britt-Louise Andersson, SIWI)

While estimates of global water withdrawals are imperfect, total human use of water from streams and other surface and groundwater bodies went up over the last century by about a factor of 8 (Shiklomanov 1993). During the same period, the population increased about fourfold. The production of goods and services, as measured by GDP, increased about 19 times at constant prices, as illustrated in Fig. 2.2. The average annual rate of growth of the global economy during the last century was 3 % (IMF 2000), which is higher than the population growth rate.

The total amount of goods and services produced during the twentieth century exceeded the cumulative total output of human history from before 1900 (IMF 2000). The longer time perspective provides even more food for thought. Discussing the links, or the lack thereof, between prosperity and growth, Jackson (2011) mentions that the global economy was 68 times bigger in 2008–2009 than it was in 1800 and that global GDP in 2100 will be 80 times larger than it was in 1950, that is, if current growth rates prevail.

For all the variables illustrated in Fig. 2.2, the most rapid increases occurred after the 1950s. But the rate of increase of population and withdrawals of water slowed toward the end of the century, while the economy continued to grow at a high rate. This may be seen as a decoupling of population and economic trends, on the one hand, and trends in water withdrawals on the other. As noted in Gleick et al. (2003; see also Chap. 8), some regions have recently seen a complete leveling off of total water withdrawals, and even, as in the case of the United States, a decrease since the late 1970s or early 1980s.

2.4.1 Water Infrastructure as "Temples of Modernization"

Construction of large hydraulic works during the twentieth century took place in various parts of the world. One of the first, massive, and complex water projects undertaken in the last century was the construction of the giant Hoover Dam in the western part of the United States. Built in the 1930s, it became the greatest single source of electricity generation in the world, with provision of water and electricity not only to nearby communities but also to distant places like Los Angeles. It was part of a vision to colonize the West and to prove that the parched desert lands could be conquered by humankind and made to bloom (Reisner 1986; Postel 1999). As with other giant projects, it had been thought about for a long time. It was possible to transmit electric power over long distances, many plans for big water projects turned into something concrete.

Globally, the drastic increase in water withdrawals after the industrial revolution started in the 1960s, as is shown in Fig. 2.2. The introduction of the Green Revolution in the 1960s, with new seeds for higher yields, required an augmented and more reliable provision of water through irrigation systems. Increased use of chemicals, fertilizers, and other inputs meant a boost to water productivity in India as well as in other countries.

The examples commented upon above refer to the exploitation of surface water. The peak in the construction of big water infrastructure programs occurred in the 1970s (WCD 2000) as environmental issues became more evident and of concern. Yet big infrastructure projects are still being undertaken. More recent projects include the Great Manmade River groundwater project in Libya, the Southern Anatolia Project in Turkey (Oclay Ünver 1997), the Three Gorges, and numerous other reservoir projects, including the South–north water transfer project in China (Gleick 2009a; Skov Andersen 2011), and the planned hydropower development in the Mekong Basin. Various countries in Africa are considered to be ripe for hydropower development in the near future. The challenge facing reservoir planners today is how to site, design, and operate them in ways that achieve the desired benefits while minimizing the social costs as well as the environmental and eco-logical degradation resulting from altered hydrologic and sediment regimes downstream.

2.4.2   Water Provision and Regulation Against Floods

Regulating and taming surface flows is vital to mitigate the risk of serious floods in many countries. For India, Lannerstad (2009) compiled information on incidences of failure of the monsoon and the related human sufferings in southern India during a 100-year period from the beginning of the nineteenth century. He noted "scarcity, desolation and disease" were rampant on 18 occasions with a duration of 1 year or 2 of consecutive years. For individual years, like 1808, failure of both monsoons "… carried off half the population." (Lannerstad 2009, 4).

With the huge population in the region, the scale of the challenge is even more severe today. For the 540 million people living in the Ganges-Brahmaputra-Meghna Basin, 50 % of the annual precipitation falls in 15 days and 90 % of the runoff occurs in 4 months (Grey and Sadoff 2007); climate change and population growth continue to worsen the situation.

2.4.3   Increasing Dependence on Groundwater

The number of new reservoirs for impounding water peaked in the 1970s (WCD 2000). The reduction in the growth rate of new reservoirs since then does not, however, mean that the demand for water has reduced. Instead, reliance on ground-water has accelerated. Today, global groundwater withdrawals are estimated to total some 650 km^3/year—about 20 % of total global withdrawals for irrigation (Wada et al. 2012). More than 70 % of that groundwater use occurs in just five coun-tries. In India, for instance, the aggregate withdrawals increased from an average of 10 to 20 km^3/year around 1950, to between 240 and 260 km^3/year in 2000, which is roughly one-third of the total global withdrawals (Shah 2009). The changeover of irrigation from surface water supply systems to groundwater systems in India has been phenomenal over the last 50 years. The use of groundwater has helped improve production because farmers can access that water when they need it rather than when the irrigation authorities decide to supply it. But it has not come without significant cost. Shah (2009, pp. 29, 187) describes groundwater extraction as "… atomistic and anarchistic." In some areas, particularly in western India where extraction exceeds the sustainable yield, a catastrophic collapse of some aquifers is likely unless better regulation and management practices can be developed quickly. Other countries with major withdrawal issues are the United States, Pakistan, China, Iran (Giordano 2009), and Libya, where groundwater supplies over 95 % of the total water demand.

2.5   Different Types of Water and Their Relative Significance

In addition to the water withdrawn from blue water sources—surface and ground-water—society depends to a large degree on rainwater stored as soil moisture. This is characterized as "green water" (see Chap. 6). Figure 2.2, therefore, only gives a partial illustration of the significance of the exploitation of water. In terms of quantity, the amount of water that is withdrawn from rivers and lakes and pumped from aquifers constitutes less than 5 % of the average precipitation over terrestrial systems. But this 5 % plays a tremendously important role in human lives and in any society. Households, industrial and other urban activities, and hydropower totally depend on this 5 %, with the largest fraction supplied to the irrigation sector. The relatively small amounts of the blue water resources are vital for human daily needs and indirectly for GDP. The much bigger green water resources are, in contrast, the blood stream for rainfed agriculture and for a number of functions in the landscape. Both blue and green water resources are derived from rainfall and their relative importance varies from one place to another and over time.

An estimated 75 % of the total freshwater withdrawals are allocated to agriculture and this sector accounts for an estimated 80–90 % of the consumptive use of this water (Foley et al. 2011). Most of the water used in agriculture or, generally, in the open landscape, returns to the atmosphere as vapor, whereas water used in the other sectors is much less consumptive. After use in industry and, generally, in indoor activities, water largely returns to rivers and other water bodies, but often downstream of where it was abstracted and of poorer quality.

Water use does, therefore, involve reliance on different categories of water and the consequences vary in terms of depletion, redistribution, and changes in the quality parameters of the resource.

2.6   Checks and Balances: Drivers That Modify Thinking and Policy

There is no doubt that the expansion of irrigation systems and the development of multipurpose water development schemes have resulted in a boost to global food production, hydropower, and urban, industrial, and rural development (CAWMA 2007).

At the same time, performance has often been below, or much below, expectations and a number of side effects have been serious. M. Reisner's popular book *Cadillac Desert* (1986) is a thorough account of how development-driven policies in the United States, which were formed during a period when it was a major concern to settle the West, have had devastating and long-term negative effects on the environment (aquatic ecosystems, pollution loads, etc.) and water availability for downstream users.

Contemporary thinking on water issues tends to be a reaction to ideas that were considered modern, progressive, and informed by science just one or two generations ago. Policies, programs, and projects that were praised in the recent past are now often regarded as a less wise manipulation of water to meet development objectives in society. Postel (1999) argues that water development technologies and related policies in the United States during the early parts of the last century, which were mimicked in other parts of the world, have generated a need for conservation methods that will protect Earth's water for the next century.

2.7 A Convenient Truth: Trend Break in Population Dynamics

Population projections have been of great public interest and concern since the days of Thomas Robert Malthus (1766–1834). Size and spatial distribution changes in populations naturally affect water demand and use. Projections about the future total size of the population are, generally, a most important piece of information. The biennial assessments of the UN Population Division (UNFPA 2011) projects three scenarios for the next few decades to 2050—nearly eight billion (low), 9.2 billion (medium), and 10.5 billion (high). Whether we will see a population peak followed by a decline or continued population growth will depend, primarily, on the assumptions about long-term fertility levels and, to a lesser degree, on mortality.

Population projections show continued growth for a few more decades. Whether the population will stabilize or start to decline after that is, however, rarely contemplated or discussed. As illustrated in Fig. 2.3, even small changes in fertility rates will produce significantly different levels of total population in the long run.

The world population passed the seven billion mark during autumn 2011—up some 0.8 or 0.9 billion in only a decade. Given the momentum of population growth and the still high fertility rates in many parts of the world, we can expect a continued increase by about 2 billion until 2050. Later in the century the population may

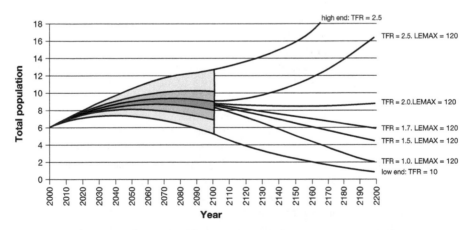

Fig. 2.3 Total world population in billions: probabilistic projections until 2100. *Yellow*, 95 % interval; *green*, 60 % interval; *blue*, 20 %; and extensions to 2200. *TFR* total fertility rate, *LEMAX* maximum life expectancy. *Source*: Lutz and Scherbov 2008

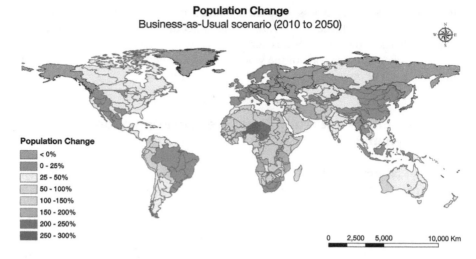

Fig. 2.4 Percentage population change 2010–2050. *Sources*: IFPRI 2010, calculations made by IWMI; Sood et al. (2013) (map prepared by Aditya Sood, IWMI)

continue to grow to between 10 and 12 billion or reach a peak and start to decline, depending mostly on the future course of fertility.

Fertility and mortality rates have decreased in several countries where, for instance, better education is available. In countries with a high population density, like Bangladesh, the number of children that would be born to a woman if she were to live to the end of her childbearing age, has dropped from about 7 to 2.3 during the last 25 years (UNFPA 2010). Fertility figures for neighboring countries, which together have about half of the world's population, have also declined. The number of children per woman is: for China, 1.6; for India, 2.7; for Pakistan, 3.5; and for Indonesia 2.1 (UNFPA 2010). These rates are much lower as compared to the situation in these countries in the recent past. Some demographers believe that fertility rates are even lower than those presented by the UN (Lutz, personal communication). In Lutz and Samir (2010) a set of long-term global projections based on scenarios covering a wide range of possible future fertility levels are presented. Selected findings are shown in Fig. 2.3.

However, fertility rates are higher in some areas that are poorly endowed with water and other resources. For instance, Chad has a fertility rate of 6.1 children per woman, Central African Republic 4.7, Congo Democratic Republic 5.9 and Congo Republic 4.6 (Fig. 2.4) (Lutz and Samir 2010). These are also areas where climate change is likely to cause increasing stresses, e.g., for agricultural production (see Chap. 6) and urban water supplies (see Chap. 5). A milestone in human history was passed in 2010, when the previous urban minority became the majority for the first time in history. Urban *per capita* water demands are generally higher than rural demands and the demand by urban residents for goods and services has direct implications for the use of water, land, and energy in rural areas.

2.8   The World Economic Map Turns Upside Down

The accelerated change and the staggering expansion of the world economy during the previous century (Fig. 2.2) were primarily associated with growth in OECD countries. Current economic prospects show that the world economic map is turning upside down (Fig. 2.5). A few distinguishing traits in this process are worthy of discussion here. The growth forecast for 2012 for high-income countries is projected to decrease while growth in sub-Saharan Africa remains robust with a rate of 4.9 % in 2011, accelerating to 5.3 % in 2012, and 5.5 % in 2013 (World Bank 2012).

Projections of economic growth rates are more uncertain as compared to demographic projections because they may change up or down in the short-run. However, the economic growth rate has generally been high over extensive periods and even accelerated in parts of the world in recent decades while there has been a gradual slow-down in the average rate of annual global population increase (UN 2004). Data on increases in population numbers and growth rates of GDP do hide important differences. In simple terms, two persons are double one person, but a doubling of GDP does not imply that purchasing power or disposable income will be twice as big. Of course, another major difference refers to the fact that it is important to recognize the implications of the social distribution of income, whereas demographic dynamics are different.

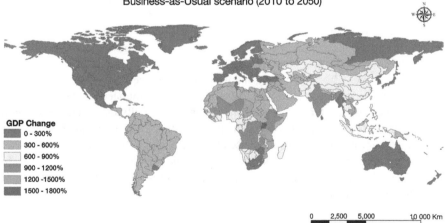

Fig. 2.5 Projections for growth of GDP 2010–2050 for the business-as-usual scenario. *Sources*: IFPRI (2010), calculations made by IWMI; see Sood et al. (2013) (map prepared by Aditya Sood, IWMI)

2.8.1 Income Distribution Determines Access to, and Mix of, Goods and Services

Even if aggregate economic growth has been rapid for many years, the uneven distribution of income means that a majority of the population, globally and in most countries, still have a low income. As a result, they have a limited access to the goods and services they need or want. These constraints hit producers, such as peasant farmers, who do not have the money to buy high quality seeds and other inputs for production. They also hit consumers with low and uncertain incomes. Even small increases in the prices of inputs for production as well as the prices for commodities may be disastrous for the poor.

More than 20 % (1.4 billion) of the world's population have incomes below a poverty line of USD 1.25 per person per day (Shah 2010). About half of the world's population has a disposable income of less than double the current poverty line, i.e., USD 2.5 per person per day. More than five billion have an average disposable income that is less than USD 10 per person per day (Shah 2010). These figures suggest that a majority of the world's population still has limited opportunities to effectively demand goods and services at a level that they might want.

For those segments of the population that are still in the low income bracket, we should expect that a considerable part of their extra income will be spent on more food, shelter, and other goods and services for basic livelihood. For people who are a bit higher up the disposable income ladder, the additional purchasing power, to a large extent, will be spent on higher quality food items, more travel, better housing, etc. Growth in GDP is likely to result not only in more commodities but also through a shift in composition in demand, toward a growing fraction of those that require higher fractions of water.

Income distribution, demographic features, and sociopolitical circumstances will largely determine the aggregate pressure on water and other resources. An increase in demand for water and other natural resources will occur in a diversified economy with stronger influences from the urban, industrial, and service sectors. As discussed in Chap. 8, these diversified economies have significantly stimulated improved resource use efficiency, while, at the same time, they have substantially increased the potential for human well-being. But improvements in GDP at national or global levels and the associated improvements in resource use efficiencies do not necessarily mean corresponding improvements in the social situation. One would expect that increases in *per capita* GDP and improved *per capita* food supply would result in a reduction of undernourished people, but, unfortunately, that is not the case globally.

2.9 Projecting Water Demands

Projections of water demand in the coming decades have been made based on various model studies carried out by various research centers (Reilly and Willenbockel, 2010). The International Food Policy Research Institute's (IFPRI) IMPACT model, the Food and Agriculture Organization of the United Nations' (FAO) World Food model, the Basic Linked System model of the International Institute for Applied Systems

Analysis (IIASA), and the International Water Management Institute's (IWMI) WATERSIM model are based on projections about population growth and the trends in GDP—a proxy for purchasing power/disposable income. The analysis for this study is based on the WATERSIM model and concentrates on three combinations of demographic change and economic growth—low population growth and high GDP growth, medium population and economic growth, and low population growth combined with high GDP growth. These three combinations of demographic and GDP trends are used to estimate the possible increases in the pressure on water.

The graphs in Fig. 2.6 show the estimated consumptive use of water in seven regions of the world from 2010 to 2050. Figure 2.7 shows the same information, but split for the main sectors of water use—irrigation, industrial (manufacturing, energy, and agro-based industry), domestic/household, and livestock. The consumptive use of water in agriculture refers to the harvested area. The estimates of consumption in Figs. 2.6 and 2.7 refer to the depletion of the water resources in terms of evaporation and transpiration. As such, they are considerably lower than the volumes of water that are withdrawn from water bodies and the amounts of water supplied or allocated, because evaporation and transpiration occur in several stages of water to harvesting throughout the growing season.

As shown in Fig. 2.7, the percentage of irrigated consumptive water use goes down substantially in 2050, especially for the optimistic scenario. The reason is related to the assumptions built into the model that there are limits concerning the amounts of water that are available in streams and other surface and groundwater bodies that can be exploited for supply to the various sectors in society. Environmental flow requirements are considered in this calculation. When aggregate demand exceeds the naturally given supply level, allocations must first be given to domestic and industrial uses. The logical consequence is that allocations to irrigated agriculture must be reduced. These findings are in line with an earlier estimate made through the same model (CAWMA 2007).

The irrigation water demand is based on the climatic conditions and the area of irrigated land. For the three scenarios presented in Figs. 2.6 and 2.7, there is no climatic change (as 30 years of historic monthly averages of hydrological inputs are used), while changes in irrigated land are limited to those due to either lack of water or price fluctuations.

The assumptions and parameters that are built into the model used to obtain these projections, and the estimates that are calculated, illustrate one possible trend in the future. Alternative assumptions would give other estimates. The validity of the assumptions and the degree of realism in the estimates arrived at are topics for debate and scrutiny.

The implications from these and other estimates on future water predicaments are several:

- In the foreseeable future, the natural availability of water will impose growing hydrological, technical, environmental, and financial constraints on society in many countries.
- With continued rapid urbanization and development of industrial and service sector activities, and with the increasing purchasing power among large segments of the world's population (most of whom will be living in urban centers) the mix

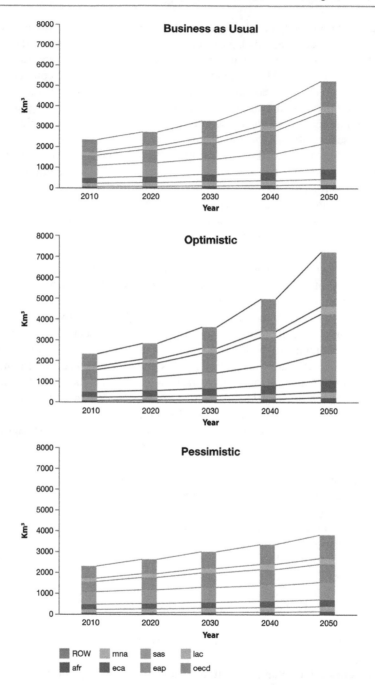

Fig. 2.6 Estimates of the consumptive use of water in seven regions 2010–2050. *BAU* business-as-usual scenario, *OPT* optimistic scenario, *PES* pessimistic scenario, *sas* South Asia, *oecd* Organisation for Economic Co-operation and Development (34 countries), *mna* Middle East and North Africa, *lac* Latin America and Caribbean, *eca* Eastern Europe and Central Asia, *eap* East Asia and the Pacific, *afr* sub-Saharan Africa (diagram prepared by Aditya Sood, IWMI)

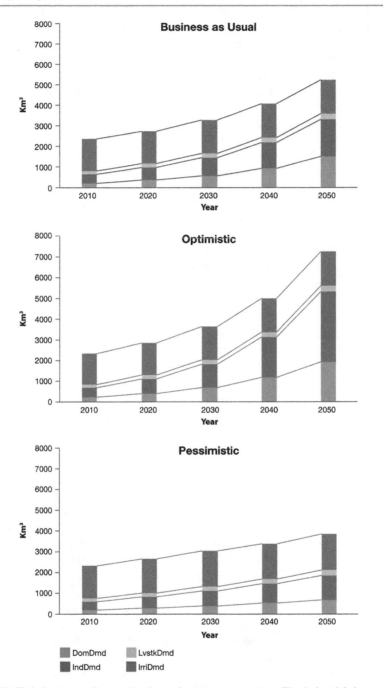

Fig. 2.7 Global consumptive use in the main water use sectors. The industrial demand also includes demand by agro-based industries. *BAU* business-as-usual scenario, *OPT* optimistic scenario, *PES* pessimistic scenario, *DomDmd* Domestic, *IndDmd* Industry, *LvstkDmd* Livestock, *IrriDmd* Irrigation. *Source*: Sood et al. (2013) (diagram prepared by Aditya Sood, IWMI)

of demands for water will change. Relative allocations of water to the urban and domestic sectors are likely to increase at the expense of allocations to the agricultural sector.

- The geographical disparities between countries that have adequate water resources and those that face development constraints because of water shortages will become acute.
- Remedies and options for mitigation and adaptation to the escalating challenges include tough changes in three major respects:
  - The actual harnessing, exploitation, and use of the various fractions of the water that are potentially available through annual precipitation must be more efficient.
  - The beneficial use of the goods and services that are more efficiently produced should increase through an improved efficiency in the supply chain, from "field to fork" (Lundqvist et al. 2008).
  - The access to basic goods and services for decent and secure livelihoods must improve for those segments of the population who are currently deprived of such an access.

2.9.1 Efficiency, Effective Goal Achievement, and Net Water Savings

The projections of the likely increases in the total consumptive use of water illustrated in Figs. 2.6 and 2.7 highlight the challenge of meeting expectations from a growing number of people who will need and demand more water, as well as goods and services that will require more water to produce. An intensified exploitation and use of water, in turn, will affect the food sector (see Chap. 6), energy sector (see Chap. 7), and our environment (see Chap. 4). With these kinds of challenges expected, and in an era of growing uncertainty and variability of water availability, it makes sense to focus on efficiency in resource use in terms of, for instance, "more crop drop," "more employment per drop," etc. Globally and in many countries, significant improvements have been made in terms of increased output per unit of water and other resources used in production, and it is important that this development continues. But improved efficiency in resource use alone does not guarantee that social and other development objectives are met. It is essential to consider two additional issues.

One refers to a tendency that increased efficiency in resource exploitation and use tends to stimulate exploitation rather than conservation of resources. A second important issue refers to the social distribution of and access to the goods and services that are produced. Improvements in the use of resources and an increase in total production do not guarantee that those who need more water and better access to goods and services will actually be the beneficiaries. Distribution of disposable income and sociopolitical circumstances will determine the access to the goods and services produced. It is, for instance, important to note that global increases in food production on a per capita basis has paralleled an increase in the number of undernourished after 1995/96 (Lundqvist 2010). What seem to be paradoxes in the relations between efficiency, resource exploitation, and social benefits mirror the rebound effect or the so-called Jevons' paradox (see Box 2.2).

Box 2.2   The Jevons' Paradox. Also Referred to as the Rebound Effect

Well over 100 years ago, Mr. W. Stanley Jevons, a British economist, made an intriguing observation (1865). He argued that increased efficiency in the energy sector will not result in reduced total energy demand and use but, on the contrary, it will stimulate energy use over time. Mr. Jevons made his observation at a time when the steam engine, developed by James Watts a 100 years earlier, in 1765, became a major driver in terms of cheap energy for a dynamic industrial revolution. The technology was not new but had been gradually developed from the seventeenth century when the original application was to pump water out of mines. But it was with expanding markets for industrial products and a cost effective exploitation of coal that triggered the process that Mr. Jevon observed.

Arguments that highlight what seems to be paradoxes naturally attract attention. Since the publication of this seminal thesis, a large number of studies with reference to the energy sector have been made on what has been referred to as the rebound effect, or Jevons paradox. Not surprisingly, the concept has been refuted but also seen as intriguing and relevant. Rather, it is surprising that this phenomenon is referred to as a paradox.

The logic of the paradox is the following. Increased efficiency generally means that it will be possible to produce more output for a given amount of input or, inversely, that a certain output can be produced with less input. This distinction is important. In the first case, potential benefits refer to the possibility to better respond to unmet needs and wants in society. The second case hints at the opportunity for resource conservation, that is, a reduction in the exploitation of resources that are required to produce a given amount of goods and services in question. The concept may thus refer to many aspects. Key issues concern prices, elasticities and how the investments that are required for efficiency improvements will be covered and by whom. Similarly, it is important to consider how benefits will be distributed in society. Issues in this regard are social and political challenges that require a strategy for progress where communities walk on two legs: a combination of natural resources and social governance. Environmental concerns must be an integral component.

With increasing production of goods and services in relation to a given amount of input of energy—in Jevon's case—employed in production, the unit cost of production will decrease, given that other costs or inputs do not increase or increase at a lower rate. If these conditions are fulfilled it will be possible,

(continued)

Box 2.2 (continued)

in principle, for the producer to sell the produce at a lower unit price as compared to the situation before efficiency improvements. If the price on the market is reduced, the individual consumer will benefit and/or more consumers may be able to pay for that product, thus increasing the demand and potential social benefit. For the producer, lower cost per unit produced is likely to stimulate her/him to expand production, assuming that there is a demand for an augmented supply and assuming s(he) is in command over additional inputs for production. Restriction in exploitation/use rights could restrict this opportunity. Increased production may be rational with a new technology for which certain investments have been made, particularly if efficiency improvements can only be achieved with increasing scale of operation. Energy savings from efficiency improvements will therefore not necessarily result in net savings. i.e., conservation but may partly or fully be used to expand total output. It may even increase total resource exploitation, i.e., the opposite of resource conservation.

Consequently, it is conceivable that efficiency improvements yield social benefits and augmented output and supply on the market in the sector where these improvements are made, but not in resource conservation. In both dimensions, policies need to be added that ensure such objectives. The paradox doesn't say anything about the social distribution of the benefits of efficiency gains.

The concept is well known among colleagues dealing with energy issues but has hardly been applied and tested with reference to water.

While policies to increase resource use efficiency have generally been successful, progress in terms of catering for basic human needs is still below expectations.

Meeting the needs of the bottom billion and balancing the dynamic wants on the demand side is a social and political challenge. To do this requires a strategy which ensures both efficient resource exploitation and use and efforts that facilitate efficient, sound, *and* fair supply chains. Although there is a decoupling of the *rates* of GDP and population growth from the *rate* of water exploitation, the exploitation of water and other natural resources will continue, even if at a slower pace as compared to earlier periods (Figs. 2.2, 2.6, and 2.7). The estimates made in other studies vary. The FAO, for instance, has estimated that the additional water that will be required to produce the 70 % more food needed by 2050, as recommended at the World Food Summit in 2009, is of the order of 5,500 km^3. Recently, this recommendation has been reduced to 60 % for the same period, i.e., between 2005/07 and 2050 (Alexandratos and Bruinsma 2012).

There are no signs of a diminishing gap between those who have and those who do not. Poverty has been reduced in many parts of the world, but not through redistribution. With about a billion people in poverty and suffering from a lack of access to a range of goods and services, more water will have to be exploited from a range of sources. Increased efficiency in the harnessing of precipitation for productive purposes is one option but it will also be necessary to withdraw and use more of surface and groundwater. As illustrated in Figs. 2.6 and 2.7, the domestic and industrial sectors will, most likely, demand much more water. From where and how will the additional water be taken? The concern for in-stream functions in rivers and other water bodies suggests that additional withdrawals of water will be subject to physical limits—when the river is dry no water can be withdrawn—and a number of restrictions.

For agriculture, there are two major opportunities. One is to increase irrigation water use efficiency as, for instance, elaborated in the comprehensive assessment of water management in agriculture (CAWMA 2007). An alternative and complementary strategy is to better use the precipitation where applicable (see, for example, Wani et al. 2011). Depending on the amount of precipitation, in terms of the fraction of water that recharges soil moisture, may involve significant risk arising from climate variability, i.e., more irregular rainfall and more rapid return flows to the atmosphere. For food security, there is a third option—improve the efficiency in the supply chain by reducing the magnitude of the losses and waste between production and what is beneficially used, i.e., what is eaten.

2.9.2   New Thinking for Opportunities to Tackle Old and New Challenges

The miserable conditions for the bottom billion cannot be explained by a general lack of water, goods, and services, but rather by a lack of access to these. In this connection, it is relevant to point at the low level of efficiency in the supply chain. An estimated 30–50 % of the food produced globally is lost, wasted, or converted (Lundqvist 2010; Parfitt and Barthel 2010; Gustavsson et al. 2011). If a large part of the produce is not beneficially used or if it is not accessible to those who need it, the efficiency and the effectiveness in the supply chain is low (Nellemann et al. 2009). It is thus important to specify what efficiency refers to; if it is confined only to production, or if it also considers the flow of the goods and services from production to end-use.

Liberal subsidies to agriculture and also to other water-dependent sectors have been motivated with reference to poverty, regional disparities, and other objectives. Public debate and policies related to water and food security have been formulated with a thrust to expand production at a rate that is faster than the population increase (e.g., the decision at the World Food Summit 2009 regarding food production increases to 2050). But while statistics support the assertion that there is enough food produced and supplied to feed everybody properly, increases in food production and supply during the last 15 years or so have not increased food security.

2.9.3   Linking Sustainable Production to Fair Access

Notions, such as integrated water resources management (IWRM), have predominantly focused on coordination between sectors and involvement of water users in the

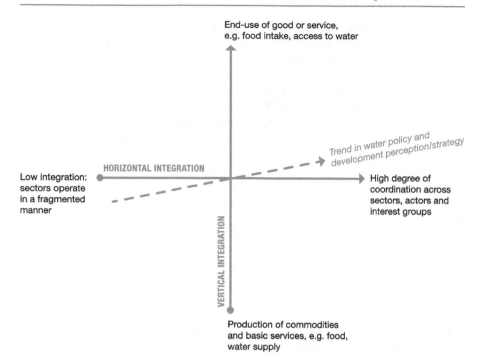

Fig. 2.8 A schematic illustration of the *horizontal* and *vertical* integration of policy and the management of resources and the goods and services produced (figure drawn by Britt-Louise Andersson, SIWI)

management and allocation of the resource between competing uses in the production of goods and services. Little, if any, attention is paid to what may be called vertical integration—from production to end-use of the produce, including the challenges associated with access.

A combination of horizontal and vertical integration is illustrated in Fig. 2.8. The arrow from the third quadrant to the first illustrates a trend in thinking and in actual strategies. Improvements have been more significant in the horizontal dimension as compared to the vertical dimension. If more attention is paid to the vertical dimension, the opportunities to provide people with goods and services can be enhanced without a corresponding increase in production. Policies directed at achieving these kinds of improvements would make it possible to meet the needs and wants in society in a more resource efficient manner.

2.10   Prospects for an Alternative "Privileged Problem Formulation"

This chapter has illustrated changes in the main drivers and the related implications for water demand and use. The purpose has also been to show that it is important to focus not only on the development and use of different types of water resources per se.

We need to pay attention to access to, and use of, the goods and services that are produced. Direct water use is, comparatively, much smaller than the indirect use, particularly water in food production. Thus it is profoundly relevant to link the figures on water use efficiency in production to information on how goods and services are distributed in society and how consumers access and use water and the goods and services produced.

A review of the main drivers and development during the period after the Second World War has been used as a reference to assess the possible and likely trends in the main drivers between now and 2050 and what might be the consequences for water use in different sectors and regions. Like other prospective studies, we recognize considerable uncertainties associated with these projections. Demographic trends are relatively more reliably predicted for the next few decades than are economic projections. Similarly, the projections about water availability, its variability, and estimates of how much of the available stocks and flows will be used are hard to make. But such projections are crucial for a better understanding of the consequences and opportunities of different courses of development.

There is a high degree of consensus in the literature that in a generation from now the world and, indeed, a large proportion of its nine billion inhabitants, are heading for a turbulent water future. Similarly, it is a common reflection that the strategies and trajectories of the past cannot continue in the same direction. Development has continuously changed course throughout history. Now and for the future, the drivers may alter the trajectory in important ways. A change of course is inevitable for the simple reason that sources of additional fresh water are literally drying up in many parts of the world. Large-scale production of water through nonconventional approaches, such as desalination, does not seem to be a viable option in the near future to meet the additional requirements pictured in Figs. 2.6 and 2.7.

One opportunity is to reduce the losses and waste in the supply chain and to improve access to the goods and services produced (Fig. 2.8). Making use of this opportunity is vital in an era when needs and wants escalate, but when water and other natural resources pose natural constraints and boundaries. The opportunities to make much better use of scarce water and other resources in *production* are thoroughly discussed in the literature and in policy. Indeed, laudable improvements have been made in this regard. No doubt, there is potential to further enhance efficiency and productivity in the use of resources. However, high levels of total production and productivity have been achieved with the help of a heavy input of fossil and other energy, high subsidies, and with a considerable effect on natural ecosystems.

Water Management in a Variable and Changing Climate

3

The world is changing, the climate is changing, and the future is uncertain. Yet we know that through our actions we can influence what our future will be. So, the question is not whether climate variability is increasing and climate change is occurring, but how can we now understand the changes and manage water for an uncertain future. Analyses of the climatic changes already observed, and the associated improvements in the methodologies we use to understand climate change and variability, have allowed us to develop improved scenarios of what the future may bring. They have also helped us to reduce the uncertainties in what we know. Because of the new technologies and our increased understanding of how to manage the risks, the remaining uncertainties can be viewed as opportunities for designing, managing, and operating our water resource infrastructure in more effective and robust ways across a wider range of possible uses and futures.

3.1 Climate Change and Water Management

Climate variability and change have important consequences for water availability, safety, and wetland ecosystems. Millions of people are affected every year by hydrological extremes, such as droughts and floods. Because of climate change the number and magnitude of these extremes is likely to increase (Bates et al. 2008). Climate change affects water availability because of changes in precipitation and evaporation, but the changes will not be consistent across all regions. In some parts of the globe, water availability will increase while in others the water available for human use and ecosystem functions is likely to decrease.

While water availability and its variability are changing with the climate, consumption by human activities continues to increase, with agriculture still, by far, the largest user of water. With continuing population growth and changing diets the amount of water required for agricultural production is likely to increase in the future. Demands from industry and for domestic use have also increased dramatically over the last decades in some regions. The requirements in most Organization

Gulbenkian Think Tank on Water and the Future of Humanity, *Water and the Future of Humanity: Revisiting Water Security*, DOI 10.1007/978-3-319-01457-9_3,
© Calouste Gulbenkian Foundation 2014

for Economic Cooperation and Development (OECD) countries have already surpassed agricultural demands and are projected to grow further. These changes in water availability and demand are causing severe water scarcity, particularly in semiarid regions, and freshwater is becoming more of a scarce good globally.

Most people involved in water resource planning are aware that as a consequence of global warming the hydrologic cycle is changing and that climate change will affect water resources management. However, it has still been difficult to use climate change scenarios for future planning purposes. How climate change will affect water availability and extreme weather events is still highly uncertain. The uncertainties make it difficult to integrate climate change information into water management.

The main purpose of this chapter is to provide information about climate variability and climate change issues that are relevant for strategic policies as well as operational practices in water management. It will offer a concise introduction to climate science and elaborate on climate scenarios, specifically the relevance and limitations of climate scenarios for water management. The focus here is on discussing recent developments in relation to climate and water. Over the last few years, improved climate scenario information has become available and the number of studies on climate change adaptation in the water sector has increased. The chapter starts with a general introduction on the effects of climate on the quantity and quality of our water resources. This is followed by a discussion on climate science and the steps in the analysis of the consequences of climate change for the water sector.

Uncertainty is one of the most important issues in climate change adaptation, so we also discuss approaches for applying information from climate change scenarios to better manage water resources in the future when the climate is both uncertain and changing. The chapter finishes with three examples of adaptive approaches. First, climate change adaptation in the Dutch Delta is discussed. Then we look at how China is integrating climate mitigation and adaptation into its water resource management plans. Finally we consider what the consequences of sea level rise and change in river runoff patterns are on saltwater intrusion in Bangladesh and the robust adaptation measures that are available there.

3.2   Effects of Climate Change on Water

3.2.1   The Earth's Energy Balance and Hydrologic Cycle

A large proportion of the solar energy reaching the Earth is used to drive the hydrological cycle. About 25 % of incoming solar energy, or about half of the solar energy that reaches the Earth's surface, is used for evaporation, producing water vapor that later falls as precipitation. The rest of the solar energy drives the oceanic and atmospheric currents, which in turn determine where the precipitation falls (Lindsey 2009). As a result of higher greenhouse gas concentrations, additional energy is available at the Earth's surface that intensifies the hydrological cycle

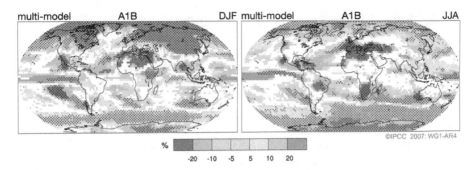

Fig. 3.1 Changes in precipitation (%) for the period 2090–2099, relative to 1980–1999. Values are multimodel averages on the Special Report on Emissions Scenarios (SRES) A1B scenario for December to February (*left*) and June to August (*right*). *White areas* are where less than 66 % of the models agree on the sign of the change and the *stippled areas* are where more than 90 % of the models agree on the sign of the change. *Source*: Fig. 1.4, IPCC (2007)

(IPCC 2007; Ludwig and Moench 2009). Climate change, then, has a large impact on precipitation patterns and, hence, changes in rainfall are expected as a result of increased greenhouse emissions. The precipitation outputs of an ensemble of climate models, illustrated in Fig. 3.1, show that some regions will receive additional rainfall while others, the subtropical regions, e.g., are likely to see less. Given the intensification of the hydrological cycle, extreme weather events, such as droughts and floods, will also become more frequent. These changes are likely to have a profound effect on human activities.

The global climate has always been variable and climate change occurs at many different scales. Temperature, rainfall, and other climate variables are continuously changing at global, regional, and local scales (Jacob and van den Hurk 2009). The recent conclusion, however, that humans are very likely affecting the global climate has initiated concern and research on how vulnerable we are to changes in the climate system and how we should adapt (IPCC 2007).

Climate variability and change at different timescales have an effect on the water sector. Four different timescales can be identified that are important for managing water resources and extremes (van den Hurk and Jacob 2009):

- The *synoptic* timescale is when individual weather events may affect hydrological extremes, such as floods.
- The *seasonal* timescale affects water management through consistently higher or lower precipitation that may result in floods, droughts, or seasonal water shortages. Seasonal weather patterns and forecasts are important for agricultural planning and management.
- At the *decadal* timescale, a projection of higher or lower than average precipitation could assist water resources planning measures, such as making adjustments to policy and infrastructure.
- Information on changes at the *multi-decadal to century* timescale can assist in designing and planning water infrastructure related to storage, water supply, and flood protection.

3.2.2 Recent Observed Changes in Temperature and Precipitation

To adapt water management to future climate variability and change, it is instructive and important to look at not only scenarios and modeling results of how the climate will change in the future but also at recent measurable changes in the local and global climates. The results presented in the last Intergovernmental Panel on Climate Change (IPCC 2012) report showed that global mean surface temperatures have been increasing since 1970. It is very likely that humans have contributed to this change.

However, for the water sector changes in precipitation are more important than changes in global average temperatures. Because rainfall is often much more variable over annual timescales than temperature, it is harder to find trends in observations. For most places on the globe statistically significant trends in rainfall cannot be found. There are some important exceptions, for example, Western Australia, northern Eurasia, and Argentina. In Western Australia, rainfall has reduced by 10–20 % over the last 30 years (Power et al. 2005; Ludwig et al. 2009). The region has a winter rainfall dominated, Mediterranean climate, and winter rainfall has been particularly affected. This has resulted in reduction in river discharges and dam inflows of more than 50 %. In the Pampas region of Argentina an opposite trend has been observed. Rainfall there has increased in some stations by more than 20 % (Asseng et al. 2012). Given the increased rainfall, land that was previously only suitable for rangeland because of limited water availability, can now also be used for crop farming. Several stations in northern Eurasia also show a clear increasing trend in precipitation, resulting in higher arctic river runoff (Adam and Lettenmaier 2008). Finally, many river systems depend on snowfall and glacial melt for significant parts of their runoff. Observations of glaciers and snow cover show trends of reduction over the last 50 years over most parts of the globe (IPCC 2007).

3.2.3 Recent Changes in Water Availability

Given the combination of human activities and climate change, water availability has changed significantly over the last decades. Population growth and economic development have increased water demand and withdrawals rapidly, particularly in Asia. Because of this rapid human development, it is often difficult to detect to what extent changes in water availability are a result of climate change or of other socioeconomic factors. Furthermore, precipitation and, hence, water resources are naturally variable in time and space. Historic surface water levels, indicated by stream flow gauge records, show large decadal and multi-decadal variations, which makes it difficult to detect the effects of climate change in these signals.

Changes in runoff are not always consistent with rainfall change, but global analyses show that in the high latitudes of both North America and Eurasia runoff is increasing. In contrast, runoff is decreasing in parts of West Africa, southern Europe, and the drier parts of South America (Milly et al. 2005; Bates et al. 2008). This is consistent with what should be expected with warmer temperatures and increased evaporation.

As a general rule, dry areas get dryer because of the greater evaporation, and wet areas get wetter because of the increased evaporation that produces greater and more intense precipitation in the areas where precipitation naturally falls. However, weather patterns also shift with changes in the climate, so this general rule is an oversimplification and is not applicable everywhere.

In addition to changes in the amount of runoff, the timing of river flows in regions with winter snowfall has significantly changed (Barnett et al. 2005). With higher temperatures, snow melts earlier in the season and during the winter more precipitation falls as rain instead of snow. The combination of earlier melt and higher winter precipitation leads to higher river discharge during (early) spring and less stream flow during the summer. Often the peak water demand is during the summer, so this change in timing of river flows can have large impacts on water resource management, causing water shortages during the summer if infrastructure and management is not adjusted accordingly.

Higher temperatures have also reduced snow cover, and glacier shrinkage has been observed around the globe (Oerlemans 2005). In Peru, for example, the area covered by glaciers has decreased by 25 % in the last 3 decades (Barnett et al. 2005). In the Andes, the disappearance of glaciers can have serious consequences for water resources because most people living west of the Andes rely on glacier-supplied river water for their water resources (Mark and Selzer 2003). Both shorter snowfall seasons and shallower snow packs have been observed here during the last decades. In the Himalayas, higher temperatures have increased runoff from melting glaciers that caused proglacial lakes to fill. When moraine walls fail, outbursts of glacial lakes and mudflows may occur. Reduction of permafrost leads to less soil stability, also increasing mudflows, rock falls, and avalanches.

In some cases reduced rainfall has caused problematic reductions in streamflows. This is especially the case in semiarid regions where small changes in rainfall can cause large changes in runoff. One of the best-documented cases is in Western Australia, where lower rainfall since the 1970s has caused large reductions in streamflow. This has reduced water availability for the Perth metropolitan area (Preston et al. 2008; Ludwig et al. 2009).

3.2.4   Hydrological Extremes

As a result of anthropogenic climate change, extreme rainfall events are predicted to increase (IPCC 2007). Extreme events are, by definition, rare and, in many regions, there is large natural variability in the occurrence of extreme rainfall events, and it is often hard to find clear trends. On a global scale, however, it is clear that the frequency of extreme rainfall events has increased over the last decades and the number of intense hurricanes (category 4 and 5) also seems to have increased over that period (Webster et al. 2005). Examples of regions where more extreme events have been observed are southern Africa (Usman and Reason 2004) and northern Australia.

The higher number of extreme rainfall events has certainly played a role in the recent increased flood frequency. The total number of floods, and the economic losses related to them, has sharply increased during the last decades (Bates et al. 2008). However, it is still unclear what the role of anthropogenic climate change has been on this higher number of floods. The authors of the Intergovernmental Panel on Climate Change's (IPCC) Fourth Assessment Report concluded that there is no conclusive evidence that the trend is related to anthropogenic climate change (IPCC 2007). The increase in flood damage is also driven by socioeconomic factors, such as concentrations of people, economic activities, and capital in vulnerable, flood prone areas. More people than ever before live in large cities that are located at the coast or near major river systems. Similar scale floods were previously much less economically disastrous than they are in the current situation. In addition, land use changes have contributed to the increased number of floods, by increasing the amount and speed of direct surface runoff. The clearing of forests and bushland increases runoff, which makes the probability of floods more likely during intense rainfall events. With the removal of mangrove forests and other coastal vegetation, natural protection against coastal floods has also been substantially reduced.

In addition to flooding, the intensity and duration of droughts have also increased since the 1970s (IPCC 2007). This is particularly the case in the tropics and subtropics. The increase in droughts is caused by a combination of reduced rainfall and higher temperatures. The Sahel region of Burkina Faso has suffered from more intense and longer droughts during the last 30 years. There are, however, some indications that rainfall has recovered in the Sahel since 1998 (Nicholson 2005). Southern and eastern Australia has become drier over the last decades and, since 2003, eastern Australia has suffered from the worst drought on record (Smith 2004). This drought has severely affected both dryland and irrigated agriculture. Many farmers have gone bankrupt and the amount of water available for agriculture has dropped dramatically. The drought has also affected industrial and domestic water supplies. Almost all the major cities in Australia have restrictions on domestic water supplies, and water companies are actively looking for new sources of water (see also the case study on Australia). Semiarid regions in North America, such as the southwestern United States and parts of southern Canada, have seen an increase in the number of droughts over the last decades as a result of declining rainfall. Population and economic growth in many areas of North America have increased the demand for water, which has made these regions much more vulnerable to droughts (IPCC 2007).

3.2.5 Future Changes in Water Availability

Possibly the most important effect of climate change is the impact it has on river discharge. Climate change has an influence on the total annual streamflow, as well as on the seasonal dynamics, for instance due to changes in the snow melt period. In general, the effects are relatively simple; higher rainfall will result in higher stream flows and reduced rainfall will decrease the stream flows. However, the correlation between changes in streamflow and rainfall are very different in different climates. In semiarid regions, river flows are highly sensitive to changes in rainfall. Generally, in

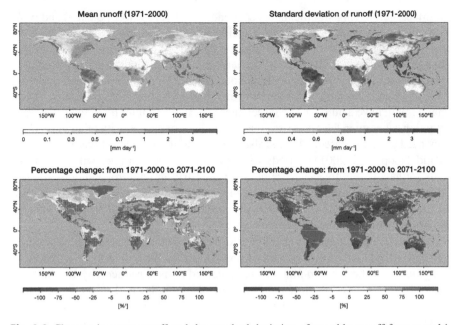

Fig. 3.2 Changes in mean runoff and the standard deviation of monthly runoff from a multi-model ensemble of global hydrologic modeling results and a multimodel global climate model ensemble with the A2 climate scenario. Hatched areas indicate no significant change. *Source*: Gudmundsson et al. (2011)

these regions only a small part of the rainfall results in runoff and most rainfall will evaporate or infiltrate into the soil. Because of this very small difference between rainfall and evaporation, a small reduction in rainfall can cause rivers to dry up completely. In Africa, in regions with an annual rainfall of less than 500 mm, a 10 % reduction in precipitation causes a 50 % lower runoff (De Wit and Stankiewicz 2006). Similarly, small increases in rainfall in areas that already flood can cause new areas to become floodplains. This is particularly true if increased rainfall comes in the form of an increase in the number of consecutive days with heavy rainfall, a likely future scenario for several regions in the world. Stream flow in many rivers in semiarid regions is already very variable, both within seasons and between years. Climate change is likely to increase rainfall variability and this will result in higher stream flow variability. Simulated changes in mean runoff and the variability of runoff between the end of the twentieth century and the end of the twenty-first century, produced using multiple climate and hydrologic models, are illustrated in Fig. 3.2.

Because a large part of society and industry depend on the freshwater surface runoff in rivers, societal effects of changes in discharge can be substantial. Existing conflicts about water appropriation—dams, irrigation, and wetland conservation— are likely to expand when water stress increases.

In colder climates a major proportion of the annual stream flow comes from snowmelt. In these regions the seasonal fluctuation in streamflow is likely to change. By 2050 in the western United States, e.g., peak flows are expected to be about 1 month earlier, having significant consequences for hydropower potential and water availability later in the year—storage facilities will be too small to retain the water that comes earlier in the season (Barnett et al. 2005). In the Rhine Basin, climate change will result in higher winter discharge as a result of intensified snowmelt and increased winter precipitation (Middelkoop et al. 2001). In the summer, discharge will be reduced because of lower snowmelt and higher evapotranspiration. These changes will have a number of consequences. To reduce future flood risks, the water retention capacity in the upstream areas and the discharge capacity of the river channels need to be increased. There is also the need to improve flood-warning systems (Middelkoop et al. 2001). Periods with low flow during summer will cause problems for navigation, for water supplies for industry, agriculture, and domestic use, and the aquatic ecosystems (e.g., fish and wetlands) that depend on the river water. At the same time, summer water demands are likely to increase because of changes in climate resulting from higher temperatures.

Almost all glaciers around the world are decreasing in size, and significant parts of all glaciers are projected to melt in the coming century. For example, the glaciers of the Tibetan plateau are projected to have reduced by 100,000 km^2 by 2035 (IPCC 2007). Half a billion people in India and 250,000 in China depend on these glaciers for their water resources (Stern 2007). The melting of glaciers initially results in increased river runoff, but will eventually cause lower stream flows when the ice has disappeared. In these cases the initial hydrological response to climate change can give a false impression of the future. This initial increase in streamflow and sudden drop later is predicted especially for the Himalayan region (Barnett et al. 2005).

3.2.6 Effects of Climate Change on Water Demand

Climate change will not only affect water resource availability and its variability but will also have an effect on water demand and use. With warmer temperatures and, therefore, greater evaporation, humans and animals will require more water. More water will also be required for domestic, energy, industrial, and agricultural uses, because of the greater cooling requirements, greater evaporation, and higher water temperatures. As discussed in more detail in Chap. 7, for example, one of the major water users in the developed world is now the power production sector, and it is rapidly using larger percentages of the water resources. Worldwide about 400 km^3 of water are withdrawn each year for the cooling of coal, gas, and nuclear-fueled power plants. In Europe and the United States the thermoelectric power sector is one of the main water users accounting for about 40 % of total water withdrawals. Climate change will have a significant impact on cooling water demands. Global warming will result in higher water temperatures, reducing the cooling capacity and, thus, more water will be needed for cooling. In addition, climate change is likely to increase electricity demands in summer as a result of the more intensive use of air conditioners. To protect river systems, there are restrictions on the amount of water that can be withdrawn and the temperatures of the waters discharged by power plants. During warm periods with relatively low river flows, conflicts often arise between environmental quality and the flow requirements of the receiving waters and the economic effects of lower electricity production (Van Vliet et al. 2012). This example shows how conflicts over water between environment and industrial production can increase as a consequence of climate change.

Agriculture (see Chap. 6) still uses the largest quantity of water globally—about 70 % of the total withdrawals. Only about 17 % of global cropland is irrigated, but this 17 % produces 40 % of the world's agricultural output (FAO 2002). Any change in agriculture resulting from climate change, then, can have a large impact on water resources, and the climate will certainly have an effect on agriculture and the water demand of the agricultural sector. Crop water demands will change with higher temperatures, but growing seasons will also get longer so that higher yielding crop varieties can be used or more crops can be grown on the same land in 1 year (multi-cropping). The consequences of climate change on irrigation water requirements in 2080 have been calculated to be 20 % above a base case scenario in which climate change is not considered (Fischer et al. 2007). But such estimates are subject to considerable uncertainty. The relationship between climate, agriculture, and water is multifaceted and complex. Climate change will change the areas that are suitable for different types of crops and, therefore, change the crop mix.

The areas where crops are grown and the types of crops that are grown will change and, therefore, change the water demands. This will also be true of natural vegetation, ecosystems, and species, which will also change as the climate changes. In many cases, higher latitudes will become more suitable for agriculture while the dry areas become less suitable climatically. Figure 3.3 provides an example. It shows the modeled changes in the suitability for rainfed wheat between the end of the twentieth and twenty-first centuries and how the suitability shifts northward.

Figure 3.3 also indicates that if the same crops continue to be grown in the same area, the crop water requirements will change as evapotranspiration changes with temperature. Even then the biological properties of plants make it difficult to predict exactly how much plant transpiration will change. Greater temperatures result in greater evapotranspiration, but greater carbon dioxide also has some fertilization effect on plants, which both speeds growth and results in the plants transpiring less water. However, the results of this carbon dioxide fertilization effect are unclear over large areas, and any gains may be more than countered by poor growth quality and, in addition, the negative effects of heat stress on plants.

Estimating water demand changes resulting from climate change in the future is further complicated by the fact that the climate changes happen as part of a larger system, and demands can be changed considerably by how we decide to use and manage land and water. Behavioral changes with respect to water use, energy use, diets, water use efficiency, technological change, and much more will all affect water use. For example, water use for energy can change considerably based on the cooling technology used; agricultural water use depends largely on what we decide to eat and, therefore, what crops are grown; and building reservoirs for water supply can result in a large percentage of the renewable resources being evaporated from reservoir surfaces.

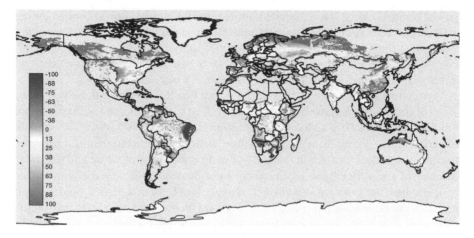

Fig. 3.3 Change in the suitability for rainfed wheat between the base period and the 2080s using the HadCM3 model with the A1F1 scenario. In *green areas*, the land becomes more suitable, and in the *red areas*, less. *Source*: Fischer et al. (2002)

Finally, water uses have a feedback on water supply and the climate system. Land use changes will change evaporation patterns, which, in turn, change weather patterns and precipitation patterns, altering the entire cycle. Climate change, then, is altering the hydrologic system in multiple ways through both supply and demand, affecting multiple sectors that are all integrated as part of a complex system.

3.3 Basics of Climate Science: What a Water Manager Needs to Know

3.3.1 Climate Change Assessment Process

The main steps of climate change assessments are essentially the same steps used in water resource management:

1. Collect observed data and information and analyze the current situation, the trends over time and the reasons for those trends
2. Develop narratives of what the future may look like in terms of the socioeconomic and biophysical structures based on an understanding of the current state and the factors that are driving changes
3. Quantify changes in the driving forces that are consistent with the narratives. In the case of climate change, this means quantifying socioeconomic and biophysical changes that alter emission levels and then calculating the resulting changes in atmospheric concentrations coinciding with these emissions levels
4. Calculate the changes associated with these changes in the driving factors, in this case the resulting climate change
5. Assess the outcomes and the management options

Depending on the situation, step five can be broken down into a number of additional steps. Since output from climate models cannot yet be used directly to assess all types of impacts, these additional steps have been necessary for water resource assessments under climate change. These steps include:

- Further definition and refinement of scenarios within the climate scenarios for the specific effects of interest
- Downscaling the climate data
- Applying additional data and models for the impact assessments

The climate and water information chain, as used by the IPCC in past reports, were based on narratives labeled A1, A2, B1, and B2, with stories of how the socio-economic system may develop in the future. The different socioeconomic assumptions result in different greenhouse gas emissions, which were quantitatively estimated based on existing information on emissions. Changing emissions result in changing atmospheric concentrations of these gases, which are quantified with the help of the models. The changing concentrations lead to climate changes in the form of changes to the energy balance (radiative forcing), leading to outcomes such as temperature and precipitation changes, sea level rise, and more, all of which are assessed qualitatively and quantitatively with models.

Errors and uncertainties are present in each step, from:

- The original data
- The large range of possible futures in the narratives
- The quantification of indicators that reflect those futures
- Errors and biases in the models themselves (climate models, downscaling methodologies, and impact models)
- The interpretation and assessment of results

The climate science community and the hydrologic community have been working to address these errors and uncertainties in each of these steps with increasing efforts. The quality and spatial resolution of climate scenarios have improved and more focus has been put on two-way communication between scientists and the users of climate data in the water sector so that the science better matches the problems that users face. Recent advances and new strategies for dealing with the uncertainties step by step are discussed here.

3.3.2 Historic Data

To analyze the effects of greenhouse gas emissions on the hydrological cycle and water management, it is necessary to develop scenarios that are descriptions of possible future pathways. But, to develop plausible scenarios of what may happen in the future we must first understand the present situation. We must also understand the historic trends in the variables of interest (drivers), such as population, economic, energy use, land use, and technology changes, and anything else that has an effect on the values we need to make decisions. This requires collecting the data, which must be done globally for the global models, a process taking many years and requiring significant resources.

Observed data can contain errors and inconsistencies from any number of sources. These include:

- Errors in the measurements themselves
- Errors in the recording of the measurements
- Changes in measurement methods and technologies used over time and between countries
- Changes in the location of the measurements
- Changes in the surrounding environment that affect the measurements
- Missing measurements, and more

Therefore, once observed data is collected, it must be analyzed, checked for errors and harmonized into consistent data sets. Fortunately, at least for the climatic data, this work has been, and continuously being, done through international efforts, and the corrected data are is available.

Additional efforts are constantly being made to make more data and information available, to critically analyze and improve existing databases further, and to indicate the uncertainty in the measured data. With regard to climate data, recent efforts combine a wide variety of measured data from various sources. These include weather stations, ships, planes, and radio observatories with satellite measurements, and mathematical weather prediction modeling of the atmospheric processes to catch inconsistencies in the data and produce more consistent and accurate datasets of climatic variables over time. This process, called reanalysis, is performed and made available by several meteorological centers internationally. These centers include:

- European Centre for Medium-Range Weather Forecasts—ERA40
- Japan Meteorological Agency—JRA-25 and JCDAS
- National Aeronautics and Space Administration-Modern Era Retrospective-Analysis for Research and Applications (MERRA)
- National Center for Environmental Prediction (NCEP), U.S. Dept. of Energy—CFSR, NCEP/DOE Reanalysis II, NCEP/NCAR Reanalysis I, NARR
- National Oceanic and Atmospheric Administration—20CR

In many cases, the reanalysis information is updated constantly and automatically using real-time, automatic, online measurements that are fed directly into the automatic data analysis and modeling programs. The reanalysis products can be, and have been, combined into ensembles (http://reanalysis.org). Comparisons among the reanalysis products and measured data provide an indication of the uncertainty in the data, as does the understanding gained from performing the data reanalysis itself. This type of uncertainty analysis is becoming more possible as large datasets are becoming more readily available. It is also recommended for use in modeling assessments, such as water resource impact assessments. Multiple model assessments are discussed further under "Modeling," to be found in Sect. 3.3.4.

The term reanalysis in this context is primarily for meteorological data. The reanalysis produces 3-hourly or 6-hourly data, but the spatial resolution is often still too coarse for sector impact assessments. Water resource assessments must take the analysis a few steps further to assess data resolution and the additional variables that are relevant in water management. Data resolution and quality have been improved by first applying binary interpolation of the reanalysis data to finer resolution grids. Elevation correction is then done to take advantage of higher resolution elevation data, since elevation affects the climatic variables. Finally, bias correction methodologies are applied, which compare the reanalysis data with the observed data and correct any systematic biases in the reanalysis data (Weedon et al. 2010).

The meteorological forcing data provide information necessary to assess climatic water availability and variability and to perform hydrologic modeling. In addition to this, many other types of biophysical and socioeconomic data and information are required to assess the current state and trends of water resources and the factors responsible. Examples include data and information on demographics, economics, land use, agriculture and forestry, energy production and use, technological change, environment and ecosystems, and cultural values and preferences. These sectors, and their interrelationships with water, are discussed in other chapters of this book. Spatially and temporally consistent data on many of these information types are still more difficult to obtain and, in many cases, contain greater uncertainty, particularly over large geographic areas and globally. But quality and availability of data are constantly improving. Spatially detailed consistent information on demographics, gross domestic product (GDP), energy, some of the major land uses, and agricultural production potentials and actual production are available. Examples of institutions providing the data include:

- Demographics—UN Population Division and International Institute for Applied Systems Analysis (IIASA)
- GDP—World Bank
- Energy—IIASA Global Energy Assessment, International Energy Agency
- Land use—Netherlands Environmental Assessment Agency's History Database of the Global Environment, Ramankutty, IIASA Global Agro-ecological Zones (GAEZ) v3.0
- Agricultural production—IIASA GAEZ v3.0, Land Use and the Global Environment lab at McGill University

3.3.3   Scenario Development

We cannot know the future, but we can use the best information and learning from the past to assist us in developing plausible scenarios of what may happen in the future. Scenarios are discussed in Chaps. 2 and 8. An example of a simple scenario is the projection, or extrapolation, of past trends into the future, perhaps with a few additional planning projections of important variables like population growth. These types of projections, which are often called business-as-usual scenarios, have often been used for planning purposes and are quite useful for short- to medium-term planning since the projections can be revised and the planning altered every few years.

However, longer term planning and large infrastructure developments require that we look further into the future, where trend extrapolations are no longer sufficient. The greatest uncertainty in managing water resources under climate change is not the data or model projections, but the future itself. The method now used to better understand the uncertainty of the future is the development of multiple plausible pathways that the future could take. These pathways are then analyzed to determine what those futures would mean for the socioeconomic and biophysical forces that drive climate change and affect water resources. The Special Report on Emission Scenarios of the IPCC (SRES), used in the fourth IPCC assessment, for example, uses four different story lines. Each story line represents a different emissions scenario stemming from a different type of future. The IPCC fifth assessment report, due in 2014, takes a slightly different approach, but still uses four primary, representative future paths for greenhouse gas concentrations, called representative concentration pathways (RCPs). Because of the time required to both develop full scenarios and run full climate models, the process is done in parallel for the fifth assessment. These RCPs are used to run climate models. At the same time, full socioeconomic scenarios are being developed that fit within the range of the RCPs and describe the possible socioeconomic pathways and feedbacks that can result in the modeled concentration pathways.

Because the driving forces are so diverse, climate experts, water resource experts, or experts in any one particular field, cannot develop these scenarios alone. It is now recognized that stakeholder groups from many different sectors and disciplines, including those that need to make planning decisions in their sectors, should be involved. In this way the range of scenarios are deemed both plausible by the many different sectors and are of interest and use to the relevant users. Experts in the different disciplines make sure that the scenarios developed are scientifically credible. They help to quantify what the scenarios would mean and communicate that information back to stakeholder groups for further refinement and improved communication and understanding. This process reduces some of the uncertainty in the future scenarios and the quantification of the scenarios. It improves the understanding of what is possible and also the understanding of the scenario uncertainties.

3.3.4 Modeling

Once qualitative scenarios have been developed, mathematical models again help to analyze the outcomes of these possible futures in a quantified way. General circulation models, otherwise known as global climate models (GCMs), have been developed to study the historic and future climate. The GCMs are mathematical representations of "the Earth System." At the most basic level, what the models are trying to do is calculate the Earth's energy balance; the amount of energy coming into and leaving the Earth and what happens to the energy within the Earth System. If more energy comes to the Earth than leaves, then the Earth gets warmer. Greenhouse gases warm the Earth by acting as a blanket and preventing heat and energy from leaving the Earth's atmosphere.

Understanding the energy balance means understanding the whole climate system, because the incoming solar energy is used to drive it. Of course, we would like to know how the changing energy balance affects the Earth System and, therefore, us. The GCMs, therefore, incorporate as many physical and biogeochemical processes of the Earth's climate system as possible to model the climate system and, hence, complete the energy balance. The GCMs are driven using different concentrations of greenhouse gases and aerosols. The future greenhouse gas concentrations are based on the emission scenarios that were developed using the different story lines. The greenhouse gas concentrations, calculated based on the emission scenarios, are then used to force the GCMs.

To test the performance of the different GCMs they are run for past time periods and compared to recorded observations. The performance of climate models has improved over the last decades and the models can simulate well large scale processes, such as those governing changes in average temperature, However there are still systematic errors, or biases, in the modeling of precipitation. Because of the large model bias, the output of GCM precipitation values cannot be used to directly force hydrological or water resources models (Hansen et al. 2006; Sharma et al. 2007; Piani et al. 2010). The biases in GCM daily precipitation cover the entire intensity spectrum—too many rain days, a bias in the mean, and an inability to reproduce the observed high precipitation events. To do proper analyses of the effects of climate change on water resources and hydrological extremes it is important to minimize these biases.

In part these biases arise because of the coarse resolution of the GCMs, which have had horizontal resolutions of 200–300 km. The GCMs are improving, both in terms of the physical processes they can model and in resolution. Recent experiments have been run at horizontal resolutions as low as between 20 and 50 km at the equator. The improvement in climate model resolution over the course of the first four IPCC assessment reports is illustrated in Fig. 3.4. Climate models are computationally intensive though, and it can presently take 1 year to run an experiment on a climate model. The number of necessary calculations is multiplied in high-resolution climate models, so testing the models and analyzing the enormous databases of the results produced takes considerable time.

The GCM physical process models and resolutions are being improved. Downscaling and bias correction methods have been developed to make the outcomes of lower resolution GCMs more applicable for global and regional impacts assessments. Methods have also been developed to better understand the uncertainties within the modeling approaches. One important method for understanding the range of uncertainty in the modeling is to compare the models' outputs not only to the observed data but also to the results from many different GCMs that are using similar input data. The range of output from the different GCMs indicates the range of uncertainty in the modeling and provides valuable information to anyone who wishes to use the climate model data. We have more confidence in areas where the results of several different models consistently agree. Multimodel ensembles can also be created, combining the results of many models to provide a single result with a range of uncertainty. One example produced using this method is shown in Fig. 3.1.

Model comparisons and multimodel ensembles alone, however, do not deal with biases (systematic errors that result in the values of climatic variables being consistently too high or too low). Nor do they improve the resolution to what is needed for many impact assessments. For this, downscaling and bias correction

Fig. 3.4 GCM horizontal
resolution improvements over
time. *FAR* first assessment
report, *SAR* second
assessment report, *TAR* third
assessment report, *AR4* fourth
assessment report. *Source*:
SPM.7, IPCC (2007)

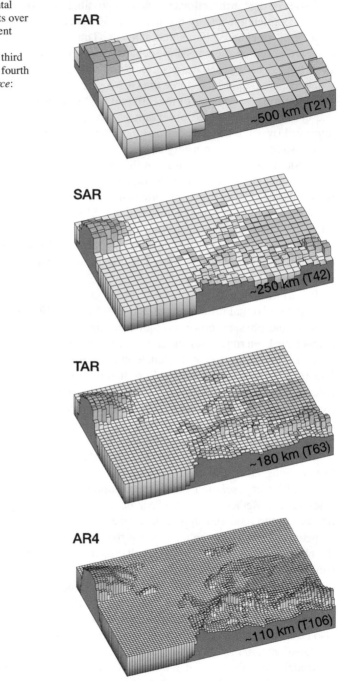

methods have been developed. The downscaling methods used can be roughly divided into two groups—statistical downscaling and dynamic downscaling (Jacob and Van den Hurk 2009). Dynamic downscaling uses high-resolution regional climate models (RCMs) nested within GCMs. The RCMs incorporate more detailed information on topographical structure, coastline, and also extensive land use information. The main shortcomings of this method are the expensive computational cost and the need for trained personnel to run and analyze the RCMs. In addition, like the GCMs that provide the boundary conditions for them, RCMs have their own uncertainties and biases that must be corrected.

Statistical downscaling techniques use correlations between large scale meteorological phenomena (which are better represented in the GCM) and locally observed values of, for example, rainfall and temperature. These correlations are then used to predict future change in local rainfall and temperature based on the modeled changes from the GCM. The advantage of this system is that it can generate future time series at the station (point) scale. Often impact models are calibrated using climate data from one or more stations. Using a statistical downscaling method it is possible to do climate change impact analyses using future climate data from the same station(s). The most problematic factor of statistical downscaling is that the correlations between different meteorological phenomena are likely to change as a result of global warming. The physical consistency between meteorological variables is not kept when using statistical downscaling. The physical consistency of a single variable spatially is also not maintained, particularly over large areas.

Neither of the techniques mentioned above are entirely suitable for global analyses, but there is increased interest in how climate change will affect the global water cycles and future water scarcity. To do global analyses on the effects of climate change on water availability, a simple "delta" method has often been used. Based on the output of a GCM, differences in temperature and relative changes in rainfall between the modeled future period and a modeled historic "base" period were calculated. These changes (deltas) were then applied to historical datasets created from observed data. The problem with this method is that it compares average rainfall and temperature changes and does not adjust for biases in the distribution of rainfall over time. With global warming, not only the average climate will change but also the distribution of rainfall and temperature. These distribution changes are important for hydrological analyses and water management because they affect drought and flood frequencies.

To overcome this problem, a combined statistical and downscaling method was developed (Piani et al. 2010; Hagemann et al. 2011). This method first interpolates the GCM output to a 0.5° by 0.5° grid and then, for each grid, applies a statistical bias correction that corrects the climate model output to produce internally consistent fields that have the same statistical intensity distribution as the observations. With this method, changes in extremes can be calculated and analyzed, a significant improvement to the standard delta method. Limitations do still exist. The method can be, and has been, used to adjust a few variables, such as precipitation and temperature, jointly for consistency between these variables, but the physical consistency among all climate variables is not entirely maintained and it is assumed that the climate model's bias is constant over time. The best way to maintain physical consistency among all climate variables is to further improve the climate models to the point where they no longer exhibit bias. In the meantime, these advances in correcting climate model biases provide the possibility to better study the impacts of not only mean changes but also changes in the probability distributions and, therefore, the extremes that are so important to understand for water management.

Once climatic variables have been appropriately downscaled and the bias has been corrected, they are used to drive hydrologic models to analyze the effect of climate change on water resources. Hydrologic modelers suffer many of the same challenges and limitations as climate modelers, including:

- Data quality and availability
- Development of realistic scenarios that reflect the interests and needs of the diverse uses of water resources and the range of possible changes in water availability and use in the future
- Adequacy of social and physical processes as they are represented by models
- Appropriate resolutions for analysis

There is always a trade-off in modeling between precision and accuracy. Adding representations of additional complex social and physical processes make the models more difficult to parameterize and run and add to the time necessary to run them. It may improve the model's precision, but without enough data on these processes, the accuracy of the model may not be improved at all or may even get worse.

The steps in managing uncertainties already described are now being used in the assessment of the impacts on water resources as well. Databases of climatic variables and socioeconomic drivers of the climate system, such as land use, population, economics, technology, and energy use change, are being improved. At the same time, databases more specific to water resources, such as river basin boundaries, runoff, water quality, groundwater aquifers, and water demands, use, and technological change in all sectors, are also being improved. Multiple scenarios are developed and compared that are consistent with the climate change scenarios, but go further to describe specifically what may happen in the future to the water variables of interest within the context of socioeconomic and climate change scenarios. Previously only experts in the water sector had developed these water-specific scenarios but now they are being developed by diverse groups of stakeholders and experts from many sectors and disciplines. This helps ensure that the scenarios incorporate the latest knowledge and information, that they are relevant and useful to all these groups, and that they include all of the feedbacks and tradeoffs in managing water for multiple needs and uses. Finally, many hydrologic models have been developed and are constantly being improved, along with mathematical representations of water use and management impacts and trends. Because many models and methods are now available, it has also become possible to do multiple model comparisons to assess the range of model uncertainties. Model comparisons have been done between the land surface models used within climate models and hydrologic models. An example of the information that can be gleaned from these comparisons is shown in Fig. 3.5.

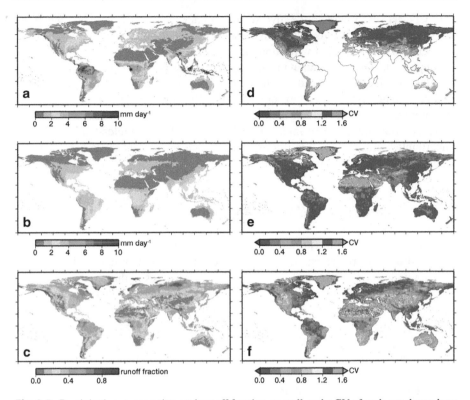

Fig. 3.5 Precipitation, evaporation, and runoff fraction, as well as the CV of each, are shown here as an ensemble created from multiple hydrologic models. The *numbers* below the maps show the range in the global average of these values from different models. All models used the same precipitation data as input, so there is only one precipitation value. Global evaporation ranged from 415 to 586 mm annually depending on the model and the runoff fraction ranged even more from 33 to 52 % of precipitation. *Source*: Haddeland et al. (2011)

Not only can changes in averages be assessed using multimodel ensembles but also because of the more advanced bias correction methods and improving model methodologies, changes in variability and extreme events can now also be assessed in more detail. Figure 3.5 shows the modeled coefficient of variation (CV) as well as the averages in a historic time period, while Fig. 3.2 in this chapter is another result of such a recently executed multimodel ensemble analysis that provides modeled future changes to averages and standard deviations.

Although we can never completely remove all uncertainty regarding climatic change and its effects on water resources, the advancements in each step of the process discussed here help to reduce the uncertainty and provide greater information about the ranges of uncertainty. Databases are being improved with additional data from more sources, as well as better data checking and analysis methodologies. Multiple scenarios are being developed and compared by larger groups of stakeholders representing decision makers and interested parties from multiple sectors and perspectives. Together, the diverse stakeholder groups can provide a wealth of knowledge and inputs and ensure that the scenarios are feasible and useful. Scenarios are being better quantified with more advanced models that represent processes at the appropriate level of detail at the appropriate scales. Model results are being carefully analyzed, corrected, and compared using downscaling, bias corrections, multimodel comparisons, and ensembles. Finally, the model results are also communicated to stakeholders who assess them to ensure they are credible and represent what the scenarios intended. All of these efforts to understand and reduce uncertainties improve the conditions under which investment decisions for water resource management must be made.

3.4   Uncertainty in Water Management

Traditionally, water infrastructure and water management systems have been designed and constructed based on historical observations of climate and hydrological data, followed by statistical analysis and interpretations of these data. For example, infrastructure is often designed to withstand an event that has a certain probability of occurring based on an analysis of the longest time series of historic data available. Infrastructure designed to withstand a 100-year flood is designed for a flood event that has a 1 % chance of occurring in any given year based on historical records. The implicit assumption in such calculations is that climate and hydrological systems behave as stationary systems, meaning that the statistical characteristics of, e.g., rainfall and discharge from a past time period where data is available, will remain the same into the future. Water engineers and managers generally understand that this is not the case, but can only work with the information they have available, sometimes adding to safety factors in the hopes of covering the additional uncertainty in future variability. Changes in climate make it even more difficult to rely on this assumption of stationarity, and historically observed data are no longer adequate to meaningfully plan for climate variability and extremes.

Water managers have been conservative about adjusting management for climate change, pointing out the big uncertainties in climate scenarios as a main obstacle to investing in adaptation measures. This has been a legitimate and understandable concern. Water infrastructure is based on the probability of certain events occurring. Managers need information about how climate change affects this probability in order to invest appropriately for the future. Figure 3.6 shows how climate information would ideally be provided. From historic data, managers can develop a probability density function (PDF), such as the solid line PDF shown in the figure. Natural and human systems have an ability to adapt to change to a certain extent with the existing knowledge and technology. These are called autonomous adaptations. Farmers, e.g., can adjust their crop mix and planting dates over time to allow for changes in climatic conditions. Other adaptations require greater investment and institutional changes. Sticking with our example, farmers may need entirely new crop varieties, new irrigation infrastructure, and new education and processing facilities once changes move beyond the ranges that can be handled by autonomous

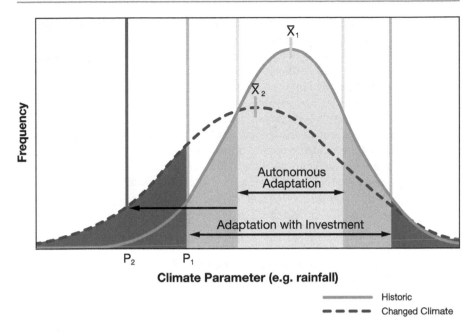

Fig. 3.6   How the probability function of a climate variable might change with climate change

adaptation. At some point, the risk may become unacceptable. In our example, this could be the point where the climate and land in an area are no longer suitable at all for agriculture.

As illustrated in Fig. 3.6, climate change can shift the entire probability distribution, changing both the mean ($\overline{X}_1$ to $\overline{X}_2$ in the figure) and the variability (indicated by the width of the bell curves in the figure). In this illustration, the probability of a drought event increases dramatically as a result of the lower average rainfall and greater variability. The probability of extreme flood events also increases somewhat because of the greater variability. The frequency of a drought of magnitude indicated by P_1 increases dramatically and a much more serious low precipitation or drought event, P_2, occurs with the same frequency. This may no longer be an acceptable risk for the current system. Of course, the level of acceptable risk could also change with new technologies. The main point is that water managers would like to know what this future probability function would look like so that they know what they will need to adapt to and plan their investments accordingly.

This is the type of information that would ideally be provided so that the proper investments in infrastructure could be made. In this case precipitation is reduced because of climate change and the risk of drought increases.

Climate models are good at modeling the large scale processes for which they were designed, and they do well at showing average changes in temperature. However, precipitation is a more local process, and GCMs have not yet been able to as accurately model changes in precipitation or its variability. Climate models alone, therefore, have not been providing this type of information at scales that are useful to water managers.

Climate modelers, hydrologists, and other impact modelers have developed a variety of methods for downscaling information from the GCMs to make them useful at river basin scales. Errors and uncertainties in each step of the climate change assessment process can make the results difficult to interpret advancements are constantly being made and the situation has improved significantly. Uncertainties are being reduced, where possible, and where not possible, more information about the uncertainties is providing more certainty through greater understanding.

Although still uncertain, changes to probability distributions under climate change can now be provided to assist water managers. In the future, these will become more and more reliable, but uncertainly can never be completely eliminated and water-planning decisions must be made. This section provides examples of how climate change information can be, and is, being used to develop more robust and flexible solutions that will provide benefits no matter what the future brings.

3.4.1 The Dutch Delta: An Example of the Possibilities for Floodplain Management

The Netherlands is a fairly small country of about 400 km^2 with more than 16 million people. It is located beside the ocean in the delta of three major European rivers, the Rhine, Meuse, and Scheldt, which are used and managed by several countries upstream. This demographic and geographic position has resulted in a long history of water resource engineering and management. Pressure on land resources has led to engineered solutions—the building of dikes and levees to reclaim land from the sea and protect against both coastal and river flooding. The result is 3,500 km of flood defenses, hundreds of locks, sluices, and pumping stations; much of the population and much of the economic activity are located in areas that are below sea level, and the entire country is dependent on the protection of the dikes.

After a major flood event in 1953, resulting in devastating economic losses and loss of life, a Delta Commission was formed that set new and very high standards for flood protection. The flood barriers at the sea were to protect against an event that should statistically happen only once every 10,000 years. These high standards, which are still valid today, are illustrated in Fig. 3.7.

After 50 years of service, however, the dikes no longer meet the necessary standards. In addition, climate change and related sea level rise are changing the probability of extreme events and, therefore, the return periods of these events. The number of people and the value of economic goods protected by the dikes have also steadily increased so that the consequences of a failure are even more serious.

The Dutch government realized that further investment in its water resource infrastructure was necessary if the Netherlands wanted to maintain its flood protection into the future. It acted by setting up another Delta Commission to provide suggestions and a plan for the long-term future of Dutch water management. The plan was to take into account the need for both flood and drought protection and also the future uncertainties arising from climate change. The new Delta Commission took the need for infrastructure renewal as an opportunity to implement water

The Netherlands – Safety Standard per Dike-ring area

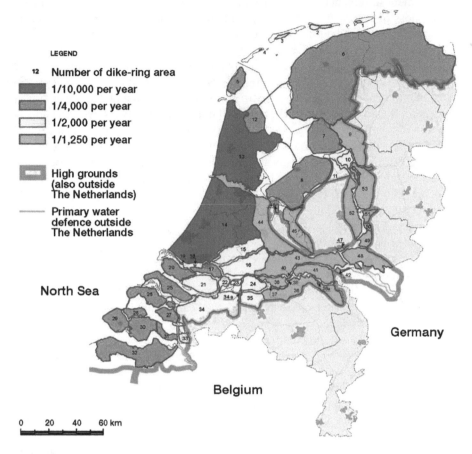

Fig. 3.7 Dike safety standards in the Netherlands. *Source*: Kabat et al. (2009)

management solutions that are more flexible, adaptable, and environmentally friendly while improving their robustness across a variety of possible future states.

Key components of the approach of the Delta Commission were to assess:

• Scenarios of global sea level rise and produce corresponding scenarios of how those translated into sea level rise locally on the Dutch coast
• Climate impacts on river flows, and then, as much as possible, to incorporate alternative and more ecologically friendly, adaptation mechanisms under the themes of "building with Nature" and making "room for the river"

The scientific studies showed that adaptation measures must deal with sea level rise and the resulting saltwater intrusion, changes in river flows, and the effects of increasing temperature and extreme events, including drought. At the same time, the measures should still supply the necessary protection and freshwater requirements for the population and economic activities.

Under the theme of "building with Nature," one of the key strategies is beach nourishment to replenish sand in coastal areas. Erosion in coastal areas contributes to loss of sand and coastal ecosystems that provide natural flood protection and dikes limit the natural river and tidal sediment deposition in these areas. Another adaptation measure is the opening of some of the dams near the sea to restore tidal dynamics, saltwater-freshwater gradients, the natural estuarine regime and ecosystems, and to improve worsening water quality. In some cases, closable floodgates will be used for times of storm surges and high river runoff to prevent flooding. The inner lake area will handle excess river flows. To support the loss of some freshwater areas, the freshwater level in one of the main freshwater lakes will be raised to correspond to the sea level rise. It is then still possible for extra water in the freshwater lake to discharge into the sea through the dam between the two. A synergy is also formed here in that power can be generated between the freshwater and saltwater and this power is potentially enough to supply the power requirements of the dikes and pumping stations of the entire water management sector (Deltacommissie 2008).

The idea behind the "room for the river" concept is to move dikes back and remove or lower barriers and structures in the floodplains to allow more room for the river to flood. At the same time, the riverbed will be lowered, excess silt buildup removed and perhaps used for beach nourishment. Wetland areas will be enhanced and polders will be made more adaptable so they can be flexibly used for flood storage and drainage as well.

Some nontraditional structural adaptation measures are also being tested and implemented. Living with floods by flood-proofing buildings has long been an option for managing flood losses. In the Dutch city of Naaldwijk, this idea is taken to the next level with the building of a floating city, which also saves energy costs by being water-cooled. Another example is the building of water storage beneath greenhouses and using these reservoirs to hold floodwater while at the same time supplying the water demand of the greenhouse (Kabat et al. 2005).

Nonstructural components of the water management plan include an adjusted political/institutional arrangement for water management with:
• Clear responsibilities for the different agencies
• A financial plan to fund the water resource development
• Coordination of river management with neighboring countries within the EU Water Framework Directive
• Authority to price water appropriately
• Responsibility for land use planning regulations

Land use regulations will require that no building shall impede river discharge or lake levels. Cost-benefit analyses are required for any new development in the floodplains and individuals and residents are responsible for their own losses in such cases.

From traditional engineering approaches, such as raising dikes, to new engineering approaches, like floating structures, to natural estuarine ecosystem enhancements, and nonstructural measures, such as water pricing, international cooperation, and land use planning and zoning, the Netherlands is implementing a flexible,

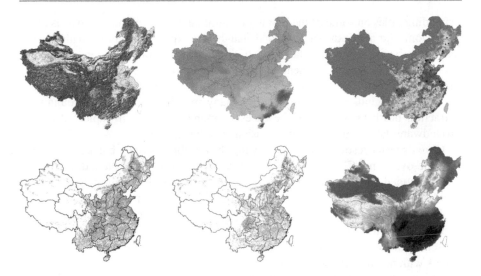

Fig. 3.8 Maps showing the spatial diversity of China. Maps show the spatial diversity of China. At the *top left* is an elevation map showing that much of China is mountainous, with *dark green* areas the only low, flat areas. The *top center* shows that the average annual precipitation varies from very dry in the northwest to very wet in the southeast. The population density map at the *top right* indicates that most of China's population lives in the low lying plains in the eastern parts of the country which are also the most suitable locations for agriculture. The map at the *lower left* shows the percentage cultivated area per square kilometer, with *dark green* indicating the highest density of cultivated area (over 80 %), moving to *yellow*, *red*, *brown*, *grey*, and *white* with progressively less cultivated area. That in the *center* shows the proportion of irrigated land in the cultivated area. The map on the *right* shows the crop water deficit, where *dark blue* indicates no deficit up to *dark red* and *violet* with more than 1,000 mm. The crop water deficit is the amount of water a crop demands beyond the amount provided by rainfall. It is the potential evapotranspiration minus the actual evapotranspiration, in this case only for a standard reference crop. *Sources*: Heilig (1999) and IIASA/FAO (2012)

adaptable, robust, and integrated water management system, employing a diversity of coordinated solutions that will sustainably provide security for its population and support for its ecosystems over the next 50–100 years. Although costly, implementing the new water management strategy now will save money when compared to the cost of reacting to hydrologic extremes when they occur.

3.4.2   China: The Challenge of Jointly Managing Great Diversity, Rapid Socioeconomic Changes, and Climate Adaptations

At the opposite end of the size spectrum is China with a land area of close to 10 million km². China's diversity and socioeconomic development present a variety of water management challenges, illustrated by Fig. 3.8. There is a precipitation gradient of from less than 50 mm per year in the high plateau and desert regions of the northwest to 2 m and more in parts of the southeast. Seventy percent of China is mountainous, only 10–15 % is arable, and over 1.3 billion people live in rapidly expanding cities on some of the best agricultural land in the plains and basins of the east. Domestic and industrial water demands are expanding rapidly with the fast pace of urbanization and industrialization. This expansion has been so fast that water infrastructure development has not been able to keep pace. Because some of these zones of rapid development are not in the water-rich areas of the country, rivers in these regions are running dry for many months of the year. Groundwater extraction has been supplying the excess demand, resulting in rapidly declining groundwater tables and saltwater intrusion into the aquifers.

China's water resources are not only highly variable spatially but are also highly variable over time. China has a long history of disastrous floods and droughts that have provided names and nicknames to China's Rivers. The Yellow River flow is highly variable, ranging from 20 to 90 billion m³/year. The river is yellow because of the sediment load it carries from variable rainfalls that erode the Loess Plateau. It has also been nicknamed "China's Sorrow" for the number of disasters it has caused. In the south, the monsoons deliver plenty of water, but also plenty of extreme events with much of the water coming in the monsoon months.

Climate change in China is expected to increase water stress mainly by increasing the variability of precipitation and runoff and more frequent floods and droughts, glacial retreat, and sea level rise. Some of these changes are already being observed. Overall, the warming trend does have a positive effect on Chinese agriculture and more areas are becoming amenable to double and triple cropping. This is a primary reason for a potential 20 % increase in production by 2050 on currently irrigated cultivated land, a result obtained using a RCM with the Agro-Ecological Zones methodology that adjusts the crop mix to use an optimal mix on that land in each time period. Unfortunately this gain in production would come at a heavy cost, as substantial changes and adaptations in the agriculture sector would be required. Crop irrigation requirements would increase by 10 % overall for the country and up to 40 % on the currently irrigated cultivated land that is also benefiting from increases in double and triple cropping.

China has a long history of water management. Although it is currently struggling to expand its water infrastructure to keep up with demand, it has seen the opportunity to also investigate climate change risks and to build a variety of flexible adaptation measures into its water resource planning. China has the world's oldest operating irrigation system, the Dujiangyan project, built over 2,000 years ago. Since that time, the focus of Chinese water management has been on structural measures to control floods and supply water. Over 83,000 reservoirs exist in China and 280,000 km of dikes, some of which have reached the end of their useful life and are at risk of failing (Zhou and Wang 2009). As a result of actions by the government to obtain scientific information and recommendations, however, the shift is now toward flexible and integrated combined structural and nonstructural approaches to dealing with climate risks. Some of the steps taken were:

- In 1988, the China Water Law established the principle of integrated water management in China, prioritized environmental concerns, and required that development be based on comprehensive scientific surveys and assessments. Water management was still undertaken at administrative levels, but some linking of the administrative levels in watersheds to produce watershed management plans was supported. Unfortunately, the law was not consistently implemented and enforced (Wouters et al. 2004).

- In 2000, the Chinese Ministry of Water Resources completed a nationwide assessment of China's water resources.
- In 2002, the 1988 China Water Law was updated. This update established stronger basin management organizations and national and regional planning; the China Water Resource Master Plan was initiated. The first stage, until 2004, was to assess the current state of water resource development, use, ecology, and environment and establish an information platform.
- In the second phase of the Master Plan, from 2005 to 2009, an action plan was drawn up concerning resource conservation, protection, allocation, and sustainable use. The Master Plan covers water issues through all sectors, lays out the development of the water resources management system, and has a focus on conservation and sustainable development in addition to security and supply. It formed the basis for integrated and adaptive management.
- In 2007, China launched the National Climate Change Program and the National Leading Group to Address Climate Change, headed by the Chinese premier, was set up to coordinate sector issues and solutions. Departments at regional levels were also required to initiate coordinating mechanisms and institutions to monitor and coordinate activities for mitigation of, and adaptation to, climate change.
- In 2010, the plan was approved by the state council and became law.
- In 2011, the Minister of Water Resources addressed the issue of coping with climate change.
- In 2011, the twelfth 5-year plan for economic development was instituted. This plan again emphasized integrated water resource management for water safety. It also stressed reduction of current and near future climate vulnerabilities and poverty related to disasters, and it enhanced public participation in water management.

The Water Resource Master Plan is the first important step toward integrated and adaptive water management. It established basin-scale water management institutions so that water management was integrated throughout entire watersheds. The plan also established mandatory controls and goals. For example, water consumption is legally limited to 670 billion m^3 in 2020 and 700 billion m^3 by 2030, which is expected to be close to the time when the population will reach its peak. Targets for ensuring a minimum volume of water being made available for ecosystems have been set for 2020 and 2030, and by 2030 water supply to the entire population will be assured even during drought. Solid targets for water use efficiency in different sectors, water quality, and water use have all been fixed.

The legal and institutional structure for integrated and adaptive water management has been set, which is already an effective adaptation solution. The focus now is on implementing the Master Plan with a variety of structural and nonstructural approaches. Since China is a large and diverse country, solutions vary from one region to another and some solutions cut across regions. Structural approaches include:

- Improvements to flood defenses
- Reservoirs
- Water treatment systems

- Advanced technologies in desalination
- Removing infrastructure from floodplains and reestablishing it in safer areas to allow rivers to flow and to flood
- Recycling water
- Harvesting rainwater

A specific example of a structural measure is the completion of three major water transfers to bring close to 45 billion m³ of water per year from the Yangtze watershed to the water scarce north. The water transfer will relieve water stress that is causing over-pumping of groundwater in the North China Plain This over-pumping has resulted in water tables dropping by 1–2 m/year in some places and in saltwater intrusion into the aquifers used to water crops. Because of environmental concerns stemming from the transfer of the water itself, more strict environmental controls are being placed on water transfers, including 260 projects and several treatment facilities to reduce pollution and ensure drinking water quality.

The nonstructural approaches being implemented have included:

- Improving flood forecasting by upgrading technology and data acquisition networks
- Continuing the successful emergency response systems and recovery services
- Flood hazard mapping
- Instituting integrated bureaus for water affairs in some cities
- Employing water saving technologies and adjusting agricultural cropping patterns to save water
- Providing ecological water allowances
- Initiating soil and water conservation and erosion control projects
- Improving water use regulations and limits
- Water pricing mechanisms
- Buying and renting land in other countries for agriculture production

Clearly, China has taken significant steps to make climate mitigation and adaptation strategies integral parts of its water resource management strategies. It will still face difficult challenges in implementation, regulation, and enforcement of many of the plans given the fast pace of socioeconomic change. However, the important steps of establishing appropriate legal and institutional frameworks for implementing robust, adaptive management systems are well underway and a variety of coordinated solutions have already begun to take shape.

3.4.3 Climate Change Adaptation in Bangladesh

Bangladesh is a very densely populated country where large parts of the population still depend on agricultural production for their livelihoods. Poverty is still widespread in the country, but the economy has been growing steadily by between 5 and 7 % each year over the last decade. Bangladesh is widely seen as one of the countries most vulnerable to climate change. The Bangladesh Delta is at the end of three large river systems, the Ganges, the Brahmaputra and the Meghna. The three rivers drain a total of about 1,200 km³ of water per year. All three rivers are strongly seasonal and about 75–80 % of this water flows down in only 5 months. Given the location in a low lying delta, in combination with very intense seasonal rainfall patterns, flood events are common and a large share of the country can be inundated with water during each event. As a result large parts of Bangladesh are flooded each year.

The dry season flows of both the Ganges and Brahmaputra are affected by the upstream construction of dams and barrages. As a result, the volume of water reaching Bangladesh in the dry season is lower than it would be under natural conditions. Climate change projections indicate a reduction of dry season flows for both the Ganges and Brahmaputra (Van Vliet et al. undated). The combination of these lower dry season flows and sea level rise is causing large scale saltwater intrusion in the southwestern coastal region. Saltwater intrusion is affecting agricultural production, urban water supplies, and the ecologically important mangrove forests of the Sundarbans.

Adaptation in Bangladesh needs to focus on the three main effects of climate change. First, higher monsoon rainfall and river flows, in combination with more frequent extreme rainfall events, are causing increased flood frequency and severity. Second, in the dry season, rainfall and river flows are reduced, causing more droughts and lower water availability. Third, the quality of the water is being reduced because of the increased saltwater intrusion.

Currently, several adaptation programs have been developed and implemented in Bangladesh. To manage the saltwater problem, more salt tolerant crop varieties are being introduced and rainwater harvesting is promoted to guarantee a freshwater supply during the dry season. In the coastal zone, aquaculture is replacing, and/or is combined with, rice farming to adapt to saltwater intrusion. Khulna, the third largest city of Bangladesh, is particularly affected by saltwater intrusion and a significant part of the population has only saltwater to drink in the dry season. To improve the urban water supply, new drinking water facilities are being developed. To reduce the saltwater problem, water will be extracted upstream where salt concentration is lower and new reservoirs will be developed to guarantee the water supply in the dry season when the saltwater intrusion is most severe (ADB 2010).

Reducing flood vulnerability in Bangladesh requires a combination of national and local scale measures. Along the coast, embankments have been developed over the last decades to reduce coastal floods. These embankments will protect the country against some sea level rise, but in the future it will be necessary to build higher and stronger embankments (Inman 2009). However, it is impossible to protect the whole country against flooding and some floods are also useful to supply water and nutrients to agricultural lands. Improved land use planning is, therefore, extremely important. Densely populated areas and land with important economic activities need high levels of protection. Land that is vulnerable to floods and difficult to protect should be set aside for agriculture or other activities, which tolerate more frequent floods. This requires land use planning at different scales. At the national scale there is an initiative to develop a Bangladesh Delta plan that will prepare the delta for different challenges in the coming century, including climate change. The Delta plan will not only look at land use planning in relation to floods but also develop a long-term vision for water management in relation to water scarcity and saltwater intrusion (Choudhury et al. 2012).

However, Bangladesh does not only need to prepare for more water but also needs to adapt to lower water availabilities in the dry season. Because of climate change, dry season flows will be reduced. This will not only cause water scarcity in Bangladesh but also in the upstream neighboring countries, in particular India. Upstream infrastructural developments in India and water extractions have already significantly reduced the dry season flow of the Ganges. This has particularly affected the flows of the Gorai River in Bangladesh. The Gorai River is the major supplier of water for the southwest region of the country. Climate change and more developments in India will further threaten the dry season flow of the river systems in southwestern Bangladesh.

One of the adaptation measures currently planned is to increase the flow of the Gorai River by either dredging the river system and/or developing barrages. This will not only guarantee agricultural and domestic water supplies but will also help to protect the Sundarbans mangrove forest, which currently suffers from salt water intrusion. It is even more important to develop a dialogue between India, China, Nepal, and Bangladesh to guarantee a sufficient water supply to Bangladesh in a changing climate.

Bangladesh has always suffered from frequent floods and disasters caused by severe cyclones, for example. Over the years communities have developed a high resilience to climate variability and the country has developed strong disaster risk reduction programs. For example, the severe cyclone Sidr, which hit the country in 2007, caused a much lower number of fatalities compared to similar events in the twentieth century. The challenges for climate change adaption are enormous in Bangladesh, but at the same time Bangladesh has one of the most advanced adaptation programs in the developing world. The main challenge for Bangladesh is to integrate adaptation in the future development programs.

3.5 The Way Forward

Planning for water management under climate change is an extension of good, integrated water resource management. We cannot know or predict the future, and the fact that our decisions and actions have an influence on the future only makes the future even more uncertain. In order to plan effectively, we must constantly be improving our knowledge and understanding and we must use the best information available to make informed decisions. Significant advances have been made in analyses and in the available information, and the work continues. We must also continue to find ways to make the existing knowledge and information, as well as examples of successful solutions, more readily available in formats that are easier to understand and apply.

However, we cannot always simply wait for more and better information before making decisions. We must plan for the future and make decisions now under existing uncertainty. We will never be one 100 % certain of what the future will bring. But our improved understanding of physical and social processes and trends, possible future changes, technologies, and management options can help us find solutions that can be effective now across a wide range of feasible future states. With the additional uncertainty arising from climate change, systems must be designed to be even more flexible and robust, incorporating a wider variety of water management techniques. Building adaptable and flexible systems now will help us make adjustments as needs change in the future. This chapter has shown how climate change can affect water resources, how uncertainties can be understood and, in some cases, reduced, and how available tools and information can be used to make informed decisions and create robust and flexible solutions to manage our uncertain future.

Water for a Healthy Environment

4

The built and natural environments in which we live can either enhance or degrade the quality of our lives. Well-designed and maintained built environments provide many of the economic and social benefits we enjoy. Nature can also provide us with multiple benefits, including products and services obtained from our natural ecosystems. These products and services include food and fuel; improved air and water quality; moderation of water flow and temperature regimes; enhanced soil formation and fertility; oxygen production; carbon and nutrient storage; recycling; and cultural, recreational, and spiritual enrichment. Water and sediment regimes within natural ecosystems are major factors in determining their health and sustainability. Withdrawals of water to meet urban demands, grow more food, and produce more energy all result in less water for the environment and for maintaining ecosystem health. Our challenge is to create a sustainable balance among all these demands that are both changing and uncertain.

4.1 Ecosystems, the Environment, and Humans

Our environment comprises everything we see, feel, breathe, taste, and hear around us. It is an ecosystem made up of multiple components that interact with each other. We humans are an integral part of our ecosystem. We are one of its components and we depend on the sustained health of all the other components for our own well-being. This fact is recognized by the United Nations (UN) General Assembly (Box 4.1). This recognition in a charter of the world's highest level governing body establishes a general principle of environmental policy making in society (Munthe 2011). But where does water fit in this recognition, and what does it imply for the manner in which we manage water resources?

Gulbenkian Think Tank on Water and the Future of Humanity, *Water and the Future of Humanity: Revisiting Water Security*, DOI 10.1007/978-3-319-01457-9_4, © Calouste Gulbenkian Foundation 2014

> **Box 4.1   Extracts from the World Charter for Nature**
>
> From the preamble:
> - Mankind is a part of nature and life depends on the uninterrupted function-ing of natural systems, which ensure the supply of energy and nutrients.
> - Civilization is rooted in nature, which has shaped human culture ... and
> - Man can alter nature and exhaust natural resources by his action or its consequences and, therefore, must fully recognize the urgency of main-taining the stability and quality of nature and of conserving natural resources.
>
> And from the general principles:
> 1. Nature shall be respected and its essential processes shall not be impaired.
> 2. The genetic viability on the earth shall not be compromised; the population levels of all life forms, wild and domesticated, must be at least sufficient for their survival, and to this end necessary habitats shall be safeguarded.
> 3. All areas of the earth, both land and sea, shall be subject to these principles of conservation; special protection shall be given to unique areas, to represen-tative samples of all the different types of ecosystems and to the habitats of rare or endangered species.
> 4. Ecosystems and organisms, as well as the land, marine, and atmospheric resources ... shall be managed to achieve and maintain optimum sustainable productivity, but not in such a way as to endanger the integrity of those other ecosystems or species with which they coexist.
>
> *Source*: United Nations General Assembly Resolution A/RES/37/7, http://www.un.org/documents/ga/res/37/a37r007.htm

4.1.1   Limits to Withdrawals of Water from Nature

All life on Earth depends on water. This has been true since the first single-cell organisms appeared some 3.5 billion years ago—consuming energy and water, growing and reproducing. From that time until very recently in geological history, there was a balance between the needs of life and the available water. Then came humans. Sometime less than 10,000 years ago we learned that we could cultivate our own food instead of just gathering it. We established civilizations and began migrating long distances. In the past 200 years our numbers have grown exponen-tially—more people to be fed, and more water demanded by each person. Today perhaps half of all available freshwater is used to satisfy human demands—twice what it was only 35 years ago (Young et al. 1994). We are not sure how much water must remain in our natural ecosystems to maintain them. However, indications are that we are approaching—and in many places surpassing—the limits of how much water we can divert from them and still preserve their health, and in turn, ours (Cosgrove and Rijsberman 2000).

Built and natural ecosystems depend on water. There are no substitutes for water. It is therefore a critically important resource that needs to be well managed, especially as it becomes scarcer and more in demand (WWAP 2009, 2012).

Water is the bloodstream of the biosphere; it links society and Nature. The amount of available freshwater in the biosphere is limited, and the demands for freshwater abstractions are increasing. In rivers, lakes, and the ground, water offers numerous benefits, including fishing, transport, tourist activities, sustaining aquatic, and riverine plant life; abstracted, it serves irrigation, energy, industry, and household uses. One could anticipate increasing conflicts between off-stream and in-stream uses of water to protect ecosystems (Falkenmark et al. 2004). This happens, but fortunately over recent years there has been improved recognition that water used to maintain the environment, or ecosystem integrity, also supports human needs by providing various services that benefit people. Studies of the role of water in ecosystems are improving our ability to value it and to understand large scale, long-term ecosystem processes and the flows of water they require (Oki et al. 2006).

Recognizing the role water plays in the biophysical and chemical processes taking place in ecosystems is an essential first step to keeping them sustainable and healthy. Inflowing water quality is as important as water quantity. Ecosystem changes may be caused by minor water quality changes. Multiple contaminants often combine synergistically to cause amplified, or different, impacts than the cumulative effects of pollutants considered separately. Continued input of contaminants can ultimately exceed an ecosystem's resilience, leading to dramatic and possibly irreversible losses. Groundwater systems are particularly vulnerable freshwater resources: once contaminated, they are difficult and costly to restore.

Floods and droughts can have a substantial impact on the ecosystems of wetlands and forests. Cycles of droughts and floods are a natural part of ecosystems and they adjust to and are influenced by them. Floods and their associated sediments can recharge natural ecosystems providing more abundant water and fertile soil for plants (including food crops). Urbanization and other land use changes, poor agricultural practices, and industrialization are among those activities that can change water quantity and quality regimes in ecosystems, and hence adversely modify ecosystems (Palaniappan et al. 2010).

4.1.2 Ecosystem Services

Ecosystems can provide essential services to humans for both life support and socioeconomic development. Healthy ecosystems in built environments provide many of the economic and social benefits we enjoy. Healthy natural ecosystems, whether terrestrial or aquatic, provide multiple benefits. Biodiversity underpins the functioning and ability of ecosystems in wetlands, streams, rivers, and lakes and their floodplains, as well as along coastal intertidal zones and estuaries, to sustain the delivery of these beneficial services (Daily 1997; MEA 2005; Boyd and Banzhal 2006). Failing to maintain adequate environmental flow regimes may adversely affect ecosystem biodiversity.

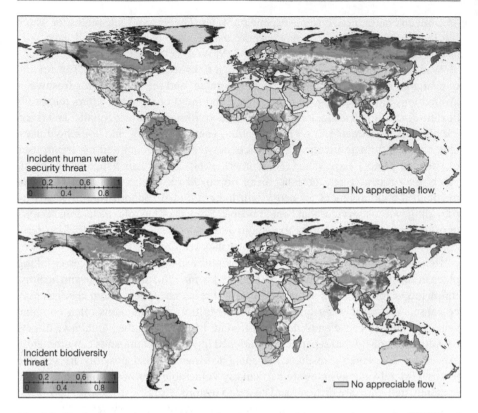

Fig. 4.1 Freshwater systems provide services in support of human water security (HWS) and biodiversity (BD) worldwide. *Source*: Vörösmarty et al. (2010)

Ecosystems, even those providing habitats for plants and animals in the desert, are genuinely water dependent. Some types of ecosystems, such as forests, bogs, or grasslands, are rainwater dependent. Others, such as certain wetlands, depend on groundwater or different mixes of freshwater and saltwater in coastal ecosystems. To protect a specific ecological service, such as for example denitrification, one needs to protect those ecosystems, in this case wetlands, that produce that type of service (Falkenmark et al. 2004).

Studies at City University of New York (CUNY) — and Conservation International provide a global-scale assessment of the ecosystem services to humans, focusing on water quality maintenance, flow stability, and flood attenuation (Vörösmarty et al. 2010). Results, illustrated in Fig. 4.1, show that freshwater systems provide these services in support of human water security (HWS). They reduce the likelihood of water shortages, and increase biodiversity (BD) worldwide. Yet threats to these services exist and call for global and site-specific remediation actions. These studies are finding population density to be an important factor in determining both HWS and biodiversity.

4.1.3 Ecosystem Services and Poverty

UN and World Bank studies suggest that the poorer a person is the more likely that person is to be dependent on food derived from his or her local environment and ecosystem. At the same time it is likely that the poorer person is involved in degrading that very ecosystem he or she is dependent upon (Shackleton et al. 2008; World Bank 2011; Brocklesby and Hinshelwood 2001).

A wide array of marine and coastal ecosystem services supporting poor people is described in the literature. They encompass all of the four types of ecosystem services defined by the Millennium Ecosystem Assessment conceptual framework (Table 4.1) (UNEP 2006).

Table 4.1 Ecosystem services from coastal and marine ecosystems that contribute to the well-being of coastal dwellers

Type of ecosystem service	Ecosystem services	Key ecosystems providing services
Supporting	• Habitat provision	• Coral reefs, mangroves, sea grass
	• Support for aquatic life cycles	• Open ocean currents
	• Hydrological cycle	• Coastal forests, wetlands, mangroves
	• Nutrient cycling	• Various coastal ecosystems
Provisioning	• Building materials (e.g., poles, limestone)	• Mangroves, coral reefs
	• Fuel (wood and charcoal)	• Mangroves, coastal forests
	• Fisheries	• All marine habitats
	• Aquaculture	• Coastal land, mangroves
	• Agricultural products	• Coastal land
	• Other natural products (e.g., honey)	• Mangroves, coastal forests
	• Employment and income	• Systems providing provisioning services
	• Medicines	• Forests, mangroves, seawater
		• Forests, coastal waterways
	• Freshwater	• Shallow lagoons
	• Seaweed production	• Coral reefs, beaches
	• Tourism income	
Regulating	• Protection from erosion	• Muddy offshore banks
	• Protection from storms and flooding	• Mangroves, coastal vegetation
	• Maintenance of air and water quality	• Mangroves, coastal forests, coral reefs
	• Waste disposal	• Open sea and tidal currents
	• Climate regulation	• Various coastal ecosystems
	• Pest and disease control	
Cultural	• Cultural identity related to coastal livelihoods	• Various coastal ecosystems
	• Education and research	
	• Bequest value	
	• Recreation	
	• Pleasant environment to live	

Source: Brown et al. (2008)

Examples of the impact of poor people on coastal ecosystem services include:

- Growing crops or permitting grazing on marginal land thereby increasing sediment loads on marine ecosystems
- Using smaller meshed fishing nets and even destructive fishing practices resulting in overfishing and habitat degradation
- Overharvesting high value or easily accessible species, e.g., sea urchins leading to cascade effects such as excessive algal growth
- Overexploiting mangrove and coastal forests for fuel wood and building materials due to lack of access to or ability to afford alternatives

Fisheries are an important component of economic wealth or natural capital (FAO 2007), and can have an important role in economic growth and poverty alleviation if appropriately managed (FAO 2005). Small-scale fisheries provide valuable protein and livelihood options to the coastal poor. An estimated 200 million people benefit from such economic opportunities through small-scale fishing in developing countries, in addition to millions for whom fisheries provide a supplemental income. Aquaculture is an increasingly important food source in Asia and to a lesser extent in East Africa (Rönnbäck et al. 2002).

4.1.4   Environmental Flows

The quantity of water needed for healthy ecosystem functioning is commonly termed *environmental flow*. This flow can have both surface water and groundwater components. Its definition is often dependent on the broad range of considerations made when implementing integrated water resources management (Box 4.2). One critical step is identifying those who will have responsibility for allocating water to competing uses (see Sect. 4.4.2) and how those decisions will be made (Dyson et al. 2004; Richter 2009).

Box 4.2   The Brisbane Declaration

At the tenth International River Symposium and Environmental Flows Conference, held in Brisbane, Australia, on September 3–6, 2007, some 750 scientists, economists, engineers, resource managers, and policy makers from over 50 nations adopted the Brisbane Declaration. Its key findings included the following:

- Freshwater ecosystems are the foundation of our social, cultural, and economic well-being.
- Freshwater ecosystems are seriously impaired and continue to degrade at alarming rates
- Water flowing to the sea is not wasted.
- Flow alteration imperils freshwater and estuarine ecosystems.

(continued)

Box 4.2 (continued)

- Environmental flow management provides the water flows needed to sustain freshwater and estuarine ecosystems in coexistence with agriculture, industry, and cities.
- Climate change intensifies the urgency.
- Progress has been made, but much more attention is needed.
  The participants committed to:
- Estimate environmental flow needs everywhere immediately
- Integrate environmental flow management into every aspect of land and water management
- Establish institutional frameworks
- Integrate water quality management
- Actively engage all stakeholders
- Implement and enforce environmental flow standards
- Identify and conserve a global network of free-flowing rivers
- Build capacity
- Learn by doing

Source: http://www.eflownet.org/download_documents/brisbane-declaration-english.pdf

Concrete measures and cases illustrating how this has been done are presented later in this chapter.

It might seem that the natural variation of flows would be needed to maintain the natural ecosystem. However, it has been demonstrated that some portion of flow can be removed without measurable degradation of the ecosystem. How much can be removed in this way is more difficult to assess. Estimates in different basins range from about 40 to 95 % of natural flow, with the natural pattern of flow also retained. Once flow manipulations move past this, river ecologists can advise on patterns and volumes of flows that will result in a range of different river conditions. This information can then be used to choose a condition that allows an acceptable balance between a desired ecosystem and other social and economic needs for water.

It is sometimes possible to restore much of the natural system, even years after no or little flow has degraded an aquatic and riverine ecosystem (Rood et al. 2003; Dyson et al. 2004; Hall et al. 2011). In other cases the damage caused by insufficient flow regimes can be irreversible.

4.2   Human Impact on the Environment

As our population grows, we have been creating an environment that in some regions is not sustainable. We are acting in ways that may disrupt or destroy ecosystems on which we are dependent. We humans are not just spreading over the planet, but are changing the way fundamental geophysical and biophysical systems work.

4.2.1   Agriculture

Agriculture is the major consumer of water worldwide. It is also a major user of land. In 2000, almost 25 % of the global land cover had been converted for cultivation. Much of this land must be irrigated. Water withdrawn from natural water bodies to grow food is water that is not available for the environment. As noted earlier, today perhaps half of economically available freshwater is being used to meet off-stream human needs. Some consequences of this are illustrated in Fig. 4.1. Box 4.3 describes those consequences arising from agriculture, together with ways to reduce them, are described in more detail by the Comprehensive Assessment of Water in Agriculture (2007). Further, agriculture, including aquaculture, is a growing source of nutrient pollution to rivers and seas.

Box 4.3   The Links Between Agriculture and Ecosystems

There are many land and water manipulations that can increase the productivity of agricultural land to meet increasing demands for more food. All have consequences for ecosystems. The key message is that agriculture makes landscape modification unavoidable, although smarter application of technology and more emphasis on ecosystem-wide sustainability could reduce many adverse impacts. These land and water manipulations include:

Shifting the distribution of plants and animals. Most apparent are the clearing of native vegetation and its replacement with seasonally or annually sown crops, and the replacement of wild animals with domestic livestock.

Coping with climate variability to secure water for crops. As water is a key material for photosynthesis, crop productivity depends intimately on securing water to ensure growth. Three different timescales need to be taken into account when considering water security: seasonal shortfalls in water availability that can be met by irrigation so that the growing season is extended and extra crops can be added; dry spells during the wet season that can be met by specific watering that can be secured, even in small-scale farming, if based on locally harvested rain; and recurrent drought that has traditionally been met by saving harvested crops, when possible, from good years to rely on during dry years.

Maintaining soil fertility. The conventional way to prevent water logging of soils, i.e., to secure enough air in the root zone, is by drainage and ditching through plowing to ensure that rain water can infiltrate. However, this can also lead to erosion and the removal of fertile soil by strong winds and heavy rain. These side effects can be limited by focusing on soil conservation actions, such as minimum tillage practices.

(continued)

Box 4.3 (continued)

Coping with crop nutrient needs. The nutrient supply of agricultural soils is often replenished through the application of manure or chemical fertilizers. Ideally, the amount added should balance the amount consumed by the crop, to limit the water-soluble surplus in the ground that may run off to rivers and lakes.

Maintaining landscape-scale interactions. When natural ecosystems are converted to agricultural systems, some ecological processes (such as species mobility and subsurface water flows) that connect parts of the landscape can be interrupted. This can affect pest cycles, pollination, nutrient cycling, and water logging and salinization. An increasing number of studies illustrate how to design landscapes to increase the productivity of agriculture while also generating other ecosystem services.

Source: Comprehensive Assessment of Water in Agriculture (2007).

4.2.2   Urbanization and Industrialization

The year 2010 marked the first time in history that over half of the world's population was living in cities. Cities are expected to continue growing and to hold 60 % of the world population in 2050 (Cohen 2010). The challenges this poses now and in the future are discussed in Chap. 5. An example is the case of China, which is rapidly urbanizing and industrializing, with associated increases in the use of energy and in the production of industrial wastes. Although economic growth from industrialization has improved quality of life indicators, it has also increased the release of chemical toxins into the environment causing in some cases environmental disasters and impacts on human health. Emissions of climate-warming greenhouse gases from energy use are rapidly increasing. Global climate change, as shown in Box 4.4 will inevitably intensify China's environmental health troubles, with potentially catastrophic outcomes from major shifts in temperature and precipitation (Xie et al. 2009). China is experiencing a growing wave of water quality and quantity challenges associated with rapid industrialization, inadequate water treatment and sewage infrastructure, weak regulatory structures for water management, and serious natural inequities in the regional distribution of water quality (Gleick 2009a).

Wherever we live, whether in rural areas or cities, we live in watersheds. In our watersheds we humans manipulate vegetation, together with the soil and water, to better meet our needs. In so doing, we often change and contaminate the landscape. The impacts of such activities on the hydrologic cycle typically exceed the impacts attributable to climate change (Vörösmarty et al. 2000). Unfortunately, taking action to mitigate the damage of such changes takes time because society has to become aware of a problem before it can consciously respond.

Box 4.4   The Environmental Impacts of Urbanization in China

China is at a stage of development at which environmental risks are changing rapidly. Over the past 30 years additional risks have been created by the largest ever migration of people from rural to urban areas. From 1978 to 2007, the proportion of people living in urban areas increased from 18 to 45 % — an increase of nearly 422 million urban dwellers (WHO-UNDP 2001; Wang et al. 2006; Brocklesby and Hinshelwood 2001). With this migration comes a complex trade-off between environmental risks in rural and urban settings. Migrants from rural areas leave behind unsafe water supplies that put them at risk of water-borne infectious diseases. But they are exposed to new risks, such as urban air pollution, exclusion from healthcare, and poor housing conditions and their related communicable diseases. Additionally, with the rising health effects of global climate change, environmental risks in China are no longer restricted to local or national emissions, but now have a substantial international component. Moreover, many other neglected issues have emerged such as occupational exposures, land use change, disposal of hazardous electronic waste, and food safety, all of which are expected to lead to increasing health problems. The scarcity of systematically collected data for these issues prohibits a credible risk analysis.

Major health effects		Populations at risk or affected
Traditional		
Unsafe drinking water and poor sanitation	Infectious diseases (e.g., diarrhea, hepatitis A, typhoid, schistosomiasis)	>40 % of rural residents (>296 million); >6.2 % urban residents (>46 million)
Modern		
Industrial water pollution	Cancers of the digestive system (e.g., stomach, liver, esophagus, or colorectal cancer)	Affected population unknown; an estimated 11 % of total digestive system cancer cases (~954,500 yearly)
Emerging		
Climate change	Deaths due to heat waves, floods, fires, and droughts; increase of infectious diseases	Throughout China, including coastal communities, water-scarce regions, and urban populations; global populations

Facing the overlap of traditional, modern, and emerging environmental dilemmas, China has committed substantial resources to environmental improvement. The country has the opportunity to address its national environmental health challenges and to assume a central role in the international effort to improve the global environment. To improve environmental health outcomes, there is a need for risk management activities that cross sectors and agencies, with a special emphasis on social and environmental inequities. A greater emphasis on regulatory enforcement is needed, to include substantial fines and criminal penalties, and systems to objectively assess the success of

(continued)

Box 4.4 (continued)

specific regulatory actions. China's recently released first National Environment and Health Action Plan emphasizes the need to coordinate activities across many sectors and ministries. It identifies coordinated and shared data collection and environmental monitoring as crucial features of future environmental health policies. Finally, successful past polices to reduce environmental risks should form the basis for reform. China has acknowledged the need to improve its environmental record, shifting its strategy from economic development to environmental sustainability in its eleventh 5-year plan, and reiterating in Copenhagen its intention to reduce greenhouse gas emissions per unit gross domestic product irrespective of the actions of other countries. These environmental goals will have a substantial effect if achieved alongside a firm commitment from the Chinese Government to formulate, implement, and enforce effective environmental health policies.

Source: Zhang et al. (2010).

4.2.3   Desertification

Some land use changes, such as deforestation, veld fires, and inappropriate farming and animal husbandry practices, can lead to degradation and desertification of watersheds and catchment areas, and reduce the amount of usable safe and clean water available downstream. Through siltation and sedimentation of rivers and reservoirs they can exacerbate soil erosion, reduce the soil water holding capacity, and decrease the recharge of groundwater and existing surface water storage capacity. The draining of wetlands reduces the availability of water that recharges groundwater, resulting in water scarcity in the long term as the groundwater table recedes. Nearly 2 billion hectares of land worldwide—an area twice the size of China—are already seriously degraded, some irreversibly. The World Water Assessment Programme (WWAP 2012) reports that almost one-quarter of the global land area was degraded between 1981 and 2003, affecting 1.5 billion people.

One of the major impacts of desertification, land degradation, and droughts (DLDD) associated with water scarcity is felt through food insecurity and starvation among affected communities, particularly in developing countries in the drylands. DLDD-related water scarcity brings about uncertainties that inevitably make communities vulnerable. If dryland countries could reduce the impacts of DLDD on water resources and achieve water security, opportunities of achieving food security would be greatly enhanced.

A variety of different measures are being applied to combat and cope with desertification. For example, in rice paddies throughout mountainous regions, particularly in Asia, terracing is employed to restrain water erosion. On less steeply sloping land, contour strip-farming works well. Conservation agriculture, including both no-till and minimum tillage, also conserves fertile soil (WWAP 2012, 120–121).

4.2.4 Coastal Zones

In some aquatic ecosystems, moderate nutrient enrichment can occasionally lead to increased populations of economically valuable fish. More severe nutrient enrichment of these same waters, however, leads to losses of catchable fish and decreases in biological diversity. The addition of nutrients such as nitrogen or phosphorous to lakes, rivers, or coastal waters with previously scarce nutrients boosts the productivity of algae (phytoplankton) that form the base of the aquatic food web. Excessive growth of algae leads to eutrophication (algal blooms) that causes a number of problems in aquatic ecosystems, particularly oxygen depletion. These changes in nutrients, light, and oxygen favor some species over others and cause shifts in the structure of phytoplankton, zooplankton, and bottom-dwelling (benthic) communities. For instance, blooms of harmful algae, such as red and brown tide organisms, become more frequent and extensive, sometimes poisoning humans who eat shellfish and killing marine mammals. Oxygen depletion can cause fish kills and create "dead zones.' Just as important, subtle changes in the plankton community and other ecological factors may trigger reduced growth and recruitment of fish species and fishery production (Howarth et al. 2002).

Even in systems where fish abundance is increased by nutrient inputs, other valued attributes, such as biological diversity, may decline. Coastal ecosystems are highly vulnerable to eutrophication so that even small increases in nutrient inputs can be quite damaging. Coral reefs and sea grass beds, for instance, are particularly susceptible to changed conditions (Lapointe et al. 2000). Major areas suffering from oxygen depletion (hypoxia), or dead zones, exist in all oceans, as shown in Fig. 4.2.

The Gulf of Mexico is an example of one of these dead zones. The seasonal reduction in dissolved oxygen (DO) occurs each year during late spring and summer following high inflows of freshwater and nutrients to the Gulf from the Mississippi River. In 2010 it covered over 22,000 km^2. Fertilizer is the major source of nitrates causing the oxygen depletion, but other sources of nitrogen in the basin include animal manure, soybeans and alfalfa, domestic effluents, atmospheric deposition, and soil nitrogen (Goolsby et al. 1997).

With most of the world's large cities located on coasts, the threat to coastal zones from nutrient enrichment will become more serious through the accelerating trend toward urbanization and the increasing discharge of human waste, often with little if any treatment, closer to the coasts.

A further threat comes from sea-level rise. The UN Development Programme (UNDP) reports that the average sea level has risen by 20 cm since 1870 and that the rate of rise is accelerating. They estimate that if the accelerated rate holds, the average level will be 31 cm higher in 2100 than in 1990. The sea-level rise in some locations will be greater. It is subject to even higher levels under conditions of high

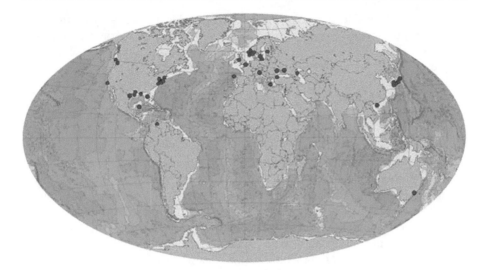

Fig. 4.2 Major coastal "dead zones." *Source*: DEAD ZONES—GES DISC Goddard Earth Sciences, Data & Information Services Center: http://daac.gsfc.nasa.gov.

tide and onshore winds. The latter can be observed to be higher in the pattern of more frequent tropical cyclones. A sea-level rise of 50 cm by 2050 would flood almost a million square kilometers and affect 170 million people (UNDP 2011). Thus the ecosystems in these coastal zones will be squeezed by urban spread and sea encroachment while at the same time receiving increasing loads of pollutants and invasion by humans and exotic species. Without corrective action, the areas of dead zones in the coastal water and ecosystems can be expected to grow. Saline surface water will move upstream in estuaries affecting the natural ecosystems located there. Freshwater aquifers will become more saline as the pressure of higher sea levels causes the migration of the saltier water.

4.2.5   What We Don't Know

Unfortunately, as these dangers increase we have inadequate knowledge about water flows and quality. The third World Water Development Report (WWAP 2009, 226) drew attention to this, pointing out that:
- Worldwide, water observation networks provide incomplete and incompatible data on water quantity and quality for managing water resources and predicting future needs—and these networks are in jeopardy of further decline. Also, no comprehensive information exists on wastewater generation and treatment and receiving water quality on a regional or global scale.
- There is little sharing of hydrologic data, due largely to limited physical access to data, policy, and security issues; lack of agreed protocols for sharing; and

commercial considerations. This hampers regional and global projects that have to build on shared data sets for scientific and applications-oriented purposes, such as seasonal regional hydrologic outlooks, forecasting, disaster warning and prevention, and integrated water resources management in transboundary basins.

- Improving water resources management requires investments in monitoring and more efficient use of existing data, including traditional ground-based observations and newer satellite-based data products. Most developed and developing countries need to give greater attention and more resources to monitoring, observations, and continual assessments of the status of water resources.

We still don't know enough about the interfaces between water and the soil and the organisms found within the water and soil. Water quality is the result of the complex interplay between physical, living and nonliving processes that ensure important ecosystem services, such as the removal of water contaminants and the provision of clean drinking water. The generation and quantitative understanding of such processes at the regional or ecosystem scale is one of the challenges of water sciences today. Data and information such as that provided by the National Rivers Water Quality Network in New Zealand (Box 4.5) will be useful for future research.

Box 4.5   New Zealand National Rivers Water Quality Network (NRWQN)

The New Zealand National Rivers Water Quality Network (NRWQN) is now in its third decade of monitoring. The NRWQN is noteworthy for being operationally stable throughout its history, and the resulting consistency is increasingly valuable for detecting water quality trends and for "anchoring" temporary special purpose monitoring campaigns. The NRWQN was carefully designed following considerable efforts to learn from monitoring experiences elsewhere. Routinely measured on 35 river systems that cumulatively drain about one half of the national landscape are: "Core" variables—conductivity, pH, temperature, dissolved oxygen, visual clarity, turbidity, colored dissolved organic matter, fecal indicator bacteria, and different forms of nitrogen and phosphorus. Associated benthic biological monitoring comprises monthly visual assessment of periphyton (algae) and annual sampling for macro-invertebrates. NRWQN data were used to provide real-world examples in a best practice guide to statistical analysis of water quality data. Data from selected sites in the NRWQN were recently used to estimate light in water bodies that is fundamental to ecological studies and also has applications to studies of fecal microbial pollutants and photochemical processes in waters. The National Institute of Water and Atmospheric Research Ltd. (NIWA) has developed a major GIS-based modeling framework with other agencies for examining the environmental and economic implications of different future

(continued)

> **Box 4.5** (continued)
>
> land use scenarios. At its core is a model for predicting nitrogen and phosphorous loads and yields at a catchment scale. The benthic biological data collected in the NRWQN have had a similar range of applications to the water quality data, ranging from scientific research to practical issues in water management. The integration of long-term benthic bio-monitoring data with hydrological and water quality data provided by NRWQN is proving increasingly valuable for informing land and water management in New Zealand.
>
> *Source*: Davies-Colley et al. (2011).

4.2.6 Tipping Points

The term *tipping point* commonly refers to a critical threshold at which a relatively small perturbation can qualitatively alter, perhaps irreversibly, the state or development of a system. Ecosystem loss and degradation may cause a tipping point where rapid and catastrophic collapse occurs following a period of apparent stability. These changes may be irreversible, requiring adaptation to new approaches to fulfilling human needs or in the worst cases with catastrophic effects for the affected human population (TEEB 2009).

Climate change can lead to tipping points. Over the past two centuries, humans have been releasing into the atmosphere fossil carbon that the planet's ecosystems had taken hundreds of millions of years to store away. Although in the carbon cycle the natural flows of carbon dioxide into and out of the atmosphere are much larger than the human contribution, the human addition unbalances those natural flows that, in turn, accelerate the effects that humans are having on water and ecosystems. These impacts include widespread changes in the distribution and intensity of precipitation, damage to coastal freshwater resources from rising sea levels, increased evapotranspiration and evaporative losses from storage areas, and accelerated melting of glaciers and icecaps.

Another example is the effect of increased use and overuse of phosphorus in agriculture. Excess phosphorus runoff from fertilized farm fields and suburban lawns enters lakes and streams. If a critical concentration is passed, it causes eutrophication and the growth of algal blooms that alter freshwater ecosystems and degrade water quality. In some instances, these blooms are toxic and pose direct threats to human and animal life. The fertilizer-fueled algae blooms themselves amplify the problem as the algae die and release accumulated phosphorus back into the water.

Carpenter and Bennett (2011) report that the "planetary boundary for freshwater eutrophication has been crossed while potential boundaries for ocean anoxic events loom in the future." Excess phosphorus in the environment is a problem primarily in the industrialized world, mainly Europe, North America and parts of Asia. Ironically, they note that soils in places like North America, where fertilizers are most commonly applied, are already loaded with the element. At the same time, phosphate deposits are found only in a few countries of the world and the material is becoming more rare. Better agricultural practices would both conserve phosphate and avoid the widespread pollution of surface waters.

Tipping points related to changes in Atlantic thermohaline circulations, the dieback and loss of Amazon rainforest, and the melting of the Greenland ice sheet receive attention in the press. In each case, some scientists believe that the gradual changes that take place in the state of these systems over time will at some point become irreversible, i.e., reach a tipping point. This can have long-term consequences for those systems (WWAP 2012).

Unlike many of humankind's other effects on the planet, the alteration of the nitrogen cycle was deliberate. In the late nineteenth century, scientists faced with a shortage of nitrogen-based fertilizer to meet global food production invented a process to manufacture it. As a consequence the annual amount of nitrogen fixed on land has increased by more than 150 % (Howarth et al. 2002). This increased agricultural productivity, but at the same time increased the volume of pollutant runoff to rivers, lakes and coastal zones as well. Algal blooms feeding on fertilizer-rich runoff waters are one result.

Today, as noted above, greater attention is being paid to the integration of knowledge and data sets to be able to better explain the interdependency of water, ecosystems, and humankind. The question is not just what are the processes involved, but also how can humans manage them.

4.3   Changing Approaches to Water Management

Our capacity to reflect, as far as is known, is unique among Earth's organisms. This makes us the only species that can knowingly destroy or protect the ecosystems — indeed the environment—in which we live. Assuming none of us want to destroy the environment of which we are part, and recognizing that some of our actions or inactions are doing just that, then we must act to create an environment that is healthy for humans.

Ecosystems have been used widely and have demonstrated their utility, particularly in moderating the variability of water quality and water flows (drought and floods). Hard engineering approaches have successfully reduced risks in rich nations, but at considerable capital and maintenance (and sometimes environmental) costs. Not all developing countries have the financial resources needed to adopt the same strategy. Nor should they have. Today even the rich countries are finding it financially difficult to continue their dependence on the use of concrete, i.e., favoring only the built environment. We are learning how to design and implement "green" strategies that take advantage of the services provided by healthy natural ecosystems together with appropriate uses of built infrastructure options even in urban habitats, to better, and more cost-effectively, manage medium- to long-term risk (WWAP 2012). This includes integrating classical methods with those of landscape ecology and recent advances in biogeochemistry at catchment scales, such as rehabilitating buffer zones, renaturalizing river channel morphology, regulating urban runoff by improved infiltration surfaces, modifying river flow regimes as a means of modifying the nutrient supply, creating wetland systems, and applying a variety of biomanipulation techniques.

4.3.1   Integrated Water Resource Management (IWRM) and Ecosystem Management

A widely accepted definition of IWRM is that of the Global Water Partnership:

> The coordinated development and management of water, land, and related resources in order to maximize economic and social welfare without compromising the sustainability of vital environmental system.

The Millennium Ecosystem Assessment (MEA 2005) recognized the value of both integrated ecosystem management and integrated river basin management. Well-managed water and watershed-based resources provide high levels of ecosystem services (IISD 2011b).

The Yellow River Basin, referred to as the Chinese mother river, is the second largest river in China. It has the highest average annual sediment transporting volume and concentration of any river in the world. Since the 1990s, it has been facing serious water problems, such as water shortage, flood threat intensification, and degradation of the ecological systems in the basin. Social and economic development in the Yellow River Basin is rapid. Since the 1950s, for instance, society, politics, and economy in China and the Yellow River Basin changed fundamentally, mainly through population growth, and economic development. About 113 million people lived in the Yellow River Basin at the end of 2007, with about 45 million in cities. The irrigated areas of the basin yielded 60 % of the total food production. Major industries, such as coal, electric power, oil, and natural gas in 2006 had output valued at about 622 billion RMB (100 billion USD), about 36 % of the GDP of the basin and 9 % of total Chinese industrial production that year. With only 2 % of the water resources of China, it supplies water to 15 % of the farmland and 12 % of the population of the country. The Yellow River Basin commission has changed its policies to manage demand and water resources to ensure sustainability (Box 4.6).

Box 4.6   Challenges of Environment Flow and Solutions Through IWRM in the Yellow River

The Yellow River Basin faced serious challenges maintaining an environment flow regime due to frequent drying-up events in the lower reaches of the Yellow River. Increasingly since 1972 the drying up of the lower reaches led to extreme cases of water shortage. In records from 1972 to 1999 there were

---

(continued)

Box 4.6 (continued)

21 years when the main courses of the lower Yellow River dried up. The most severe situation was in 1997, when there were 330 days without water flowing through the river estuary outlet into the Bo Sea. The maximum length of the river affected was 704 km in 1997. This is about 90 % of the total length of the lower Yellow River.

To balance the multiple uses of water for agriculture and industry while maintaining ecosystem health, the Yellow River Conservancy Commission (YRCC) changed its policy from water supply to water demand management using integrated water resources management (IWRM). YRCC's integrated water allocation program balances water availability with social, economic, and ecological developments. The improved, reliable supply of water has enhanced the quality of life for over 100 million people both in the basin and in regions served by the river. Large areas of wetlands and biodiversity in the Yellow River Delta have been restored, returning life and vitality to the river. Regulating water and sediment flow has significantly reduced the risk of devastating floods—protecting the 90 million people living in the flood prone areas downstream. With nine provinces and regions along the 5,464 km river, YRCC adopts a consultative approach to secure the support of the provincial governments and the people to equitably allocate water for domestic, industrial, and agricultural uses. Remote sensing and automation is employed to collect real-time river system information to monitor and control the reservoirs. Emergency response plans manage, control, and mitigate major water pollution incidents. Instead of the frequent periods of drying up of the 1990s, the Yellow River has had continuous flow throughout each year since 1999.

4.3.2 Win-Win Approaches

With the introduction of the concept of ecosystem goods and services as development assets, ecosystem health becomes an important concern in relation to poverty, hunger, water supply, and environmental sustainability. Rising concern for environmental sustainability also increases pressure to provide water to maintain ecosystems. Degradation of water quality in many basins coupled with rising standards of quality for water use in all sectors increase the importance of protecting natural water sources and the need to find better and cheaper water and wastewater treatment technologies.

The management of Lake Kinneret in Israel is an example (Markel 2005). It is the only large surface water source in Israel. With an area of 167 km^2 it supplies some 30 % of the country's freshwater. The watershed of the lake is 2,730 km^2 in area, and is intensively used for agriculture and tourism purposes. The pollution from the watershed and the necessity of pumping water from the lake are threatening the water quality of the lake. Monitoring water quantities and qualities in the lake and its watershed provides the basis for planning and decision making regarding engineering and other alternatives for managing the lake and watershed. Since 1999 some structural, technical, and logistical changes have been introduced into the monitoring-management systems. These changes led to a lake-watershed integrated monitoring system. A Monitoring Task Force coordinates the monitoring work of all organizations and guides improvements of the monitoring system. A series of management activities in the watershed reduces the nutrient loading from the watershed to the lake.

Water quality issues related to environmental health are beginning to make ecosystem health an important socioeconomic issue even in the poorest countries. Borrowing strategies from the realm of high finance, a trust fund has been created to protect rivers and watersheds in Colombia—and to help provide clean drinking water to Bogotá, the country's capital (Box 4.7).

Box 4.7 A Win-Win Solution: Bogotá Conservation Trust Fund

The fund will attract voluntary contributions from Bogotá's water treatment facilities to subsidize conservation projects—from strengthening protected areas to creating incentives for ecologically sustainable cattle ranching—that will keep sedimentation and runoff out of the region's rivers. Without such projects, the facilities have to spend millions to remove pollutants from the livestock to provide clean drinking water for Bogotá's eight million residents. Water treatment facilities in Bogotá could save USD four million per year by proactively investing in watershed protection, according to a study undertaken by The Nature Conservancy and its partners. "The best part of this initiative is that it is voluntary and will not raise water costs for residents," explains Samuel Moreno Rojas, mayor of Bogotá.

Protecting upper watersheds will also help preserve key ecosystems that provide habitat for many endangered species, including spectacled bears, Andean condors, and rare Andean guans.

The Conservancy brought together a broad range of public and private stakeholders—many of whom had never collaborated before—to navigate complex regulations and broker the landmark agreement. The fund will be managed by a board, which will include representatives from the Conservancy, the water companies and other stakeholders, including Bavaria, a brewery, and one of the biggest private companies in Colombia. Bavaria donated USD 150,000 to start up the fund. Conservancy donors in the United States provided the seed money that enabled in-country staff to launch the funds. The Conservancy will work to grow the Bogotá water fund by securing additional investments from national agencies. The Conservancy plans to implement six more such funds in South American countries in the next 2 years.

Source: http://www.nature.org/ourinitiatives/regions/southamerica/colombia/howwework/water-fund-bogota.xml

Flooding is another environmental issue. At considerable costs, engineered works are often used to protect lands and development against flooding. The benefits of such measures have indeed been substantial, but occasionally they do fail. When that happens the economic and social damages can motivate the construction of even more flood protection works, at least for a while. But it is often possible to use the natural environment to supplement the physical facilities. For example, wetlands on floodplains can serve as buffers against flooding. In London, dikes protecting a large population are now to be removed, thereby restoring wetlands that can absorb floodwaters. Massive infrastructure maintenance costs and flood insurance premiums for city inhabitants are reduced in the process (London Rivers Action Plan: http://www.therrc.co.uk/lrap/lplan.pdf). The same applies to smaller communities as well (Wilkinson et al. 2010).

Lake Winnipeg in Canada is the tenth largest freshwater lake in the world. Its drainage basin is the second largest in North America; only that of the Mississippi exceeds it. The basin is home to six million people, encompasses about 55 million hectares of agricultural land, and supports 17 million livestock and an agricultural industry of about CAD 20 billion per year. Its ratio of watershed area to lake surface area is the highest of any freshwater lake in the world, making its loadings of nutrients comparable to that of lakes in more densely populated regions. Measured by summer chlorophyll levels it is the most eutrophied large lake in the world (Lake Winnipeg Stewardship Board 2006). The Red River, flowing north from the United States, contributes as much as 66 % of the total phosphorous load and 43 % of the nitrogen to the lake. At its entry to the lake it flows through the Netley-Libau Marsh, one of the largest freshwater coastal wetlands in Canada. Unfortunately, the marsh's natural capacity to provide nutrient removal benefits has been severely compromised by drainage, dredging, flooding, and other landscape changes over the past century.

Since 2006, the University of Manitoba and Ducks Unlimited Canada have collaborated on an innovative bioeconomy project. In the pilot project tall marsh plants with slender leaves (that grow across North America) are harvested, compressed, and burned for bioenergy. The benefits of this approach are nutrient removal, bioenergy production, possible carbon credits, phosphorous recovery, and habitat improvement (Cicek et al. 2006). Depending on the economic and environmental circumstances for each application, the successful results in the Netley-Libau Marsh are replicable in other coastal zones in the Lake Winnipeg basin. The result is improved water quality and wetland health and community economic development. Thus farmers in the basin may continue to be prosperous while sustaining vibrant ecosystems. A network of stakeholders in the immediate catchment has developed in support of this approach (IISD 2011a).

4.3.3 Engineered Ecosystems

Almost everyone currently living on this planet depends on engineered infrastructure. We live and work and travel and play in engineered structures. We benefit from all the services our built environment provides. Considering just water, these services include supplies of clean water, flood protection, and the control, collection and management of storm water runoff, wastewater removal and treatment, recreation facilities, and flow augmentation for in-stream quality management. Today much of our built environment largely ignores the services our natural environment can provide. But we are learning how to combine both built and natural infrastructure, often at much lower costs than engineered alternatives, to create more desirable living and working environments for us all, especially for those living in urban areas.

Creative infrastructure development can also offer services that were not there in the natural state. Ways are being found to design, operate, and maintain systems that minimize negative externalities at a lower cost than traditional investments (Fay and Toman 2010). Public policies can offer incentives for private sector decisions regarding investment and consumption that reflect the social benefits of environmental sustainability and the costs of various forms of environmental protection. At a global level there is a need to increase environmental research and development (R&D) and encourage the international transfer of greener technologies.

The *green economy agenda* is a response to these trends and seeks to reinforce and accelerate the progress of sustainable development. It involves public policy, individual and collective business initiatives, and private customer behavior. The agenda has serious implications for water infrastructure. It increases pressure for more efficient use of resources and reductions in waste and greenhouse gas emissions, both of which aim to shift investment and consumption patterns toward alternatives that deplete fewer natural resources.

Box 4.8 illustrates how a business decision, initially motivated by profit and the need to access natural resources for production, has helped to reduce risks and uncertainties related to future water scarcity by providing an additional water reserve for communities and the environment.

Box 4.8 Restoring Water Provision in a Dry Area: Italcementi

The Sitapuram limestone mine is a captive mechanized opencast mine operated by Zuari Cement Ltd (part of the Italian Italcementi Group). It is located at Dondapadu, in Nalgonda District in southeastern India. The area sustains agriculture while two perennial streams flow through the existing mining lease area and eventually into Dondapadu Village.

The company's objective, after removing limestone, was to convert the excavated area into a lake (75–80 % of the mining area) and then develop a recreation site around it. The company also opted to develop a green belt around the lake to maintain the soil and help protect the flora and fauna.

The conversion of the excavated area into a lake included the creation of small ponds and larger water bodies, in addition to monitoring water quality and the water table. Catchment drains were constructed and connected to pits to arrest silt and sediment flowing out of the mining area. This helped reduce uncertainty by creating water reserves and decreased the potential pollution from the mining activities.

(continued)

Box 4.8 (continued)

The quarry has been operational since 1986. The adjacent green belt was developed in 2000. Bushes were planted on the slope of the pit to retain soil and protect the pit's walls from collapsing. The developed green belt along the boundary of the mining lease area acts as a barrier, protecting the surrounding area from the dust and noise created by mining activities. Jatropha plants (for biodiesel) are being grown on 9 ha in and around the mining lease areas. PVC pipelines provide a permanent water source to the trees from the quarry bench.

The results have been as follows:

- The creation of a large body of water that has attracted many birds from other areas, including ducks, cranes, and hornbills, and sometimes kingfishers if fish have spawned in water reservoirs. This adds to the preservation of the ecological environment. The reservoir also benefits the local communities that often face water scarcity and can use the reservoir for agricultural irrigation and fish cultivation;
- The recharging of the underlying aquifer, which has raised the water table in the surrounding area and increased vegetation;
- Monitoring and management of silt deposition, which prevents overflow of sediments from the mine area into the surroundings and consequent disturbance of local flora and fauna. Some of the mined pits may fill up over a longer period of time; and
- The creation of greenery around the mine premises, retention of earth due to the plantation of trees and bushes, and reduction of CO_2 levels in the atmosphere.

Source: WWAP (2012).

In the public sector, the City of Dollard des Ormeaux (Québec, Canada) provides a small example of possible action. In the late 1960s, faced with the need to build a large diameter drain to carry storm water from the rapidly expanding town, the City Council purchased a large piece of the area and deepened the lowest lying land to make an artificial lake—for about the same cost as that of building the drain. The surface runoff would go to this lake. A weir regulates the flow out of the lake and maintains a minimum level. Instead of a storm drain, this far-sighted council can be proud of the 48-hectare nature park they created. It features the lake, a forest, and playing fields used today by people in the summer for jogging and biking on trails and for picnics. Skiing, skating, and tobogganing are winter activities. It has also attracted wildlife the year round. On the negative side, algal growth arises during the summer months from storm water polluted by fertilizer runoff from suburban lawns. William J. Cosgrove Centennial Park is named after the individual who proposed its existence.

At a national level The Four Rivers Project in Korea illustrates how water resource management can become a driver for green growth (Box 4.9).

Box 4.9   Korea's Four Rivers Restoration Project

The Four Rivers Project illustrates how water resource management can become a driver for green growth. Following an economic crisis, Korea decided to allocate 2 % of its GDP per year over the 2009–2013 period (totaling USD 86 billion) on green investment with a view to solving short-term economic problems and creating jobs. Twenty percent of the green budget (USD 17.6 billion) was invested in the water sector through the Four Rivers Restoration Project (4RR). This project involves four ministries (Environment; Food, agriculture, forestry, and fisheries; Culture, sports, and tourism; and Public administration and security). The aims are to secure water supply (1.3 billion m^3); manage floods and droughts; improve water quality (BOD 3 mg/L) and restore water ecosystems (223 restoration projects); develop riverbanks to ensure space for leisure; and develop the regions around the rivers. The project, now 95 % complete, was carried out by dredging 570 million m^3 of sedimentary soils, constructing 16 new small-sized multipurpose dams and reservoirs, and removing pollutants, including in farm areas.

The government expects this project to generate USD 32.8 billion worth of positive economic benefits and create 340,000 jobs. Ultimately, the government expects the experiences and technologies developed in the project to make Korea one of the leading countries in the water management sector while contributing to its national economy.

Source: OECD (2012). *Environmental Outlook to 2050.*

Other opportunities will present themselves to engineers to take better account of the needs of, and benefits from, natural ecosystems in the process of adapting to climate change. For example, in many places, in preparation for increased precipitation it will become necessary to replace drainage conduits and culverts under barriers, such as highways and railways. While in the process of enlarging them engineers can consider the possibility of setting them at levels to facilitate the passage of aquatic species, thus perhaps reestablishing habitats that had been divided when the original works were constructed. Another possibility arises when it becomes necessary to reconsider the operating rules governing releases of water from reservoirs in response to changed patterns of rainfall. Releasing flows to correspond to the natural peak flows to which the aquatic ecosystems were accustomed will help to preserve or renew species diversity.

4.4   Valuing and Allocating Water and Ecosystems

Goal 7 of the UN Millennium Development Goals (MDGs) is to "integrate the principles of sustainable development into country policies and programmes and reverse the loss of environmental resources." Freshwater is the key environmental resource.

When the cost of building infrastructure to capture, treat, and transport this water to users exceeds what they or the country can afford, this is known as an economic scarcity. Political factors may also restrict society's ability to expand physical infrastructure to store and access additional water. The amount that can be withdrawn or consumed may be increasingly limited by new policies that set minimum standards for how much water should remain in the environment.

Water entitlements, water allocation, water distribution, and water use are dynamically linked and depend on (changing) water availability. Water issues related to environmental health have traditionally been considered a concern primarily for developed countries and their environmental movements. However, the increasing realization of the multitude of benefits of ecosystem goods and services—such as, providing food and fiber to the poorest—is gradually making ecosystem health an important socioeconomic goal even in the poorest countries (Cosgrove and Tropp 2013).

Valuing the benefits of ecosystem services and quantifying them in monetary terms can help decision makers as they balance the costs and benefits of alternative options for managing water and ecosystems. It is not always easy to do this in a way that all stakeholders accept. But it is easier when it is evident that reversing degradation and preserving the structure and the functioning of many water-related ecosystems is not only a sound ecological alternative but also economically beneficial. It is a means to show the extent of ecosystem service benefits that are foregone when water is withdrawn for other uses and to identify the potential gains attainable from reversing degradation trends. Financial instruments, such as buying land to protect it or compensating other users for not engaging in activities that can degrade ecosystems, can be used to align individual incentives to the collective interest of preserving the values supported by the natural environment.

Failure to value fully all the benefits that water and ecosystems provide through their different uses is a root cause of the political neglect of water and its mismanagement. It leads to an insufficient appreciation of the importance of both water and ecosystems, suboptimal levels of investment in constructed and natural water infrastructure, and to related policy being afforded low priority in country development programs and poverty reduction strategies.

The need to balance various perspectives in valuing water is buttressed by the Ministerial Declaration of the second World Water Summit at The Hague in 2002 that stated:

> Valuing water is to manage water in a way that reflects its *economic*, social, environmental and cultural values for its uses, and to move toward pricing water services to reflect the cost of their provision. This approach should take account of the need for equity and the basic needs of the poor and vulnerable.

Recognizing that water resources have a value, governments have begun charging companies for the water they use. In Brazil, a charge of USD 6.50/GWh represents the value attributed to the water used to produce energy. The revenue is shared between the different levels of government (Braga et al. 2009).

In Québec, the power company Hydro-Québec deducts CAD 0.62/GWh from its net revenues to be paid to a fund to aid future generations through reduction of government debt (Generations Fund) (Ben Mabrouk 2008). The forecast revenue from this source in the financial year 2011–2012 is CAD 608 million (Québec 2012).

One of the major problems in valuing water in its different uses, and especially for sustaining ecosystem benefits, is how to resolve the different perspectives of value. As Doorn and Dicke (2012) point out: "In water management, some values are represented by stakeholders, but some values do not have a representative or council. Yet, these values are essential to current and future water management." Moreover, the various benefits derived from water cannot easily be expressed in common terms, such as money, to inform economists as they identify water allocation policies that maximize social welfare. Deriving the economic value of water in its different uses is controversial, has high data requirements, is complex, and requires technical and economic skills. Different valuation techniques are applied for different uses and policy purposes (Wilson and Carpenter 1999). One of the problems has been that most cost-benefit analyses based on present value calculations discount future economic benefits with the result that they seem much lower than present costs. Lowering the discount rate is not the answer. More research is needed to adequately deal with these and other issues related to the economics of sustainable development (Wilde 2010).

The base for integrating ecosystem sustainability principles into water resource planning, development, and management already exists. Implementation presents a major social challenge. It will require substantial political will and sustained efforts from the technical community to devise water management strategies that meet both human and ecosystem needs (O'Keeffe 2009).

4.4.1 A Green Economy

Most of our domestic and international political institutions are not well suited to giving high priority to the complex issues of sustainable development, including water management. As illustrated in Chap. 7, management of energy is interlocked with management of water. Fortunately, measures to improve management in one sector may improve performance in another as well. A World Energy Council publication proposed a number of measures that would apply equally well to water management:

- A readiness to live in greater harmony both with each other, including the poorest and most deprived in world society, and with our natural environment and the other species that inhabit it.
- Advance the ethical basis of governance, economic transactions and human interaction especially in respect of open governance; publicly available and independent audits of economic transactions and their environmental impacts; and common rules for best business practices, safety and environmental performance (World Energy Council 2001).

The UN General Assembly recognized that global actions enhance national level responses. They set the eighth goal of the MDGs to "develop a global partnership for development." Institutional arrangements are necessary at the international level to address environmental challenges that are common, i.e., all nations experience them. They are also required to provide a global response when they are connected to, say, global trade or climate change, which cannot be dealt with solely at the national or basin level.

A hopeful sign is the recognition in preparation for the Rio+20 Summit that water use efficiency and demand management should be essential elements of green growth strategies. The Organisation for Economic Co-operation and Development (OECD) has indicated a number of policy orientations that can reinforce this, including "soft" infrastructure that can include natural wetlands, assessing the value of water services including "environmental services," and more (Box 4.10; OECD 2012).

Box 4.10 OECD Policy Orientations for Green Growth

Several policy orientations can more systematically harness water management for growth. They include:

- Investing in water storage and water distribution systems in water-scarce regions. Resource reliability is essential for green growth strategies. However, water storage technologies and infrastructures can disturb ecosystem balances. Soft infrastructures (e.g., wetlands, flood plains, and groundwater recharge), small-scale dams, rainwater harvesting, or appropriately designed infrastructures may be a response.
- Assessing the value of water, identifying the beneficiaries, and mechanisms to ensure beneficiaries contribute to cover the costs of the benefits they enjoy. Putting a sustainable price on water and water-related services is an effective way to signal the value of the resource and to manage demand.
- Being prepared to allocate water where it adds most value. The aim should be to avoid rents in water use, by reflecting efficiency gains in water allocations, and by diverting water to value-adding activities (including environmental services). This may trigger a reallocation among water users (e.g., from farmers to cities), a difficult policy challenge. Some OECD countries are gaining experience with regulations (water abstraction licenses which reflect scarcity), market mechanisms (e.g., tradable water rights) and information-based instruments (smart metering) in this regard. Such tools can be effective when the environmental and social perspectives are appropriately weighted, in conjunction with economic value.
- Invest in water supply and sanitation infrastructure, in particular in urban slums where unsafe water and lack of sanitation generates huge health costs and lost opportunities to the economy.

Source: OECD (2012).

4.4.2   Trade-Offs

Trade-offs between satisfying increased human demands for water and reducing the adverse impacts on ecosystems services from that water use are unavoidable and they may require that society balances social benefits to be satisfied and environmental services to be preserved. For example, water is an essential element in energy production (e.g., for biofuels, hydropower, and cooling techniques for thermal and nuclear power plants). Energy is a critical input to transfer water and to tap alternative sources of water (e.g., desalinization). In an increasing number of locations, there is competition between food and energy commodities for limited water resources. The conversion of natural vegetation to arable land to expand food production, often one of the most immediate needs of poor people, may result in the loss or decline of a number of other ecosystem services related to biodiversity and land cover, and will be a source of conflict in several regions. The building of a dam for water storage may increase the supply of freshwater when it is needed, but could disrupt some of the hydrological processes necessary for a healthy river, as well as flood productive lands and cultural sites. The effects of such trade-offs may be felt in areas distant from the original site or may be displaced in time and so are not immediately obvious. Furthermore, the poor are often "losers" in the trade-off process, particularly because they lack the power to oppose them.

For example, both mitigating and adapting to climate change might be facilitated by restoration of agricultural land in flood zones through planting trees with resulting reduction in impact of floods, improved water quality, restoration of biodiversity, and additional sequestration of greenhouse gases (OECD 2010). Thus it is possible to develop trade-offs that create positive synergies so that the total level of services supplied is improved. Adaptive management and use of scenarios can improve the assessment, monitoring, and learning processes that inform the decisions to be made (Comprehensive Assessment of Water Management in Agriculture 2007). Developing such trade-offs, especially under uncertain conditions and unknown futures, should be based on scientifically informed arguments.

4.5   Who Speaks for Mother Nature?

Earth gives us everything that is essential to life: the air, water, food, warmth, the seasons... and its innumerable wonders. If we mistreat and abuse her, she will cease to be our friend. If we do not know how to equitably distribute the gifts from her generous hands, she will not be our friend. If by creative and respectful efforts we don't keep her green, pure, and flowing with life, how will we be able to leave a friend for future generations? We won't, and future generations will raise their fists against us. The Earth is like a mother. She explores, understands, protects, and makes her produce like an immense treasure. If we make her our slave, she will not be our friend. (Jean-Yves Garneau, *Pour que la terre soit notre amie* (The Earth is our friend), 2012. (Loosely translated from the original French.))

The need for such an approach was recognized in Rio in June 2012. Articles 39 and 40 of its declaration state:

> 39. We recognize that the planet Earth and its ecosystems are our home and that Mother Earth is a common expression in a number of countries and regions and we note that some countries recognize the rights of nature in the context of the promotion of sustainable development. We are convinced that in order to achieve a just balance among the economic, social, and environment needs of present and future generations, it is necessary to promote harmony with nature.

> 40. We call for holistic and integrated approaches to sustainable development, which will guide humanity to live in harmony with nature and lead to efforts to restore the health and integrity of the Earth's ecosystem.

Ecosystems are now increasingly seen as solutions to water problems and not just their victim. There is increasing recognition of the services delivered by ecosystems, their value, and the necessity to sustain them. This is often accompanied by an increasing willingness to find solutions. While this adds another dimension to the competition among potential uses for water, it more properly reflects progress toward more balanced sustainable water management. It reinforces the need for a common framework for dialogue and consensus among multiple interests in water.

Effective stakeholder engagement is necessary to make water allocation decisions transparent and fair (Cosgrove 2008). Issues to be discussed must go beyond the sanctioned discourse: politically correct and limited to the general understanding of the audience. Improving the information available to water users is crucial if they are to make the "right" decision. Routine stakeholder consultation is not the same as true empowerment, where a community takes control of its water management, leading to more legitimate and cost-effective solutions with better chances for implementation. The unique capacity of humans to analyze and reflect on themselves and their environment means that all those sitting around the table have a moral obligation to speak for "Mother Nature."

Jack Moss and colleagues argue that "the complexity of the interfaces between many different stakeholders and the tendency for water to raise strong emotions frequently leads 'value differences' to become 'value divides'" (Moss et al. 2003). These can end in polarization that blocks dialogue and prevents reasonable governance solutions. Improved understanding of the value differences can help identify commonalities and interdependencies that may be useful in negotiated agreements.

4.6 Meeting Human Needs on a Sustainable Planet

As this chapter has emphasized, humans are linked to and dependent on the ecosystem of which they are a part. That system has always been subject to natural change. Whether consciously (manufacture of nitrogen) or unconsciously (pollution), and whether for good or evil, we humans are changing our environment faster and apparently in different directions than it would change following its natural course. Some of the factors contributing to this are:

- Exponential population growth and increasing per capita consumption
- Domestication of plants and animals

- Industrial products and processes disrupting the key biogeochemical cycles

In many places the ecosystem, including the hydrological cycle, can no longer meet the demands of humans who occupy it. This condition is spreading day by day.

Fortunately, hope also lies with humankind. By renewing our sense of unity with the rest of Nature, we can use the unique collective intelligence we have developed over millennia to create a sustainable future for the lifetime of the planet. We can imagine new ways of being and through cooperation and innovation we can achieve them. This is the message that 50,000 stakeholders from every domain of civil society, representing both genders, and from across generations brought to The Future We Want conference in Rio in June 2012.

Water can be an element that encourages cooperation to achieve this objective. The water cycle links humans and landscapes, on the scale of individual species, local ecosystems, and the global ecosystem (Falkenmark et al. 2004). Dooge et al. in the preface to the COMEST Report on the Ethics of Freshwater Use, support this concept: "Somehow water forces us to go deeper than the familiar adversarial positions and acknowledge what we really share—a respect for life and well-being. Water can be a superordinate value, the appeal to which is capable of coalescing conflicting interests and facilitating consensus building within and among societies. In a sense, negotiations over water use, itself, could be seen as a secular and ecumenical ritual of reconciliation and creativity" (UNESCO 2003).

Unfortunately, until now development has been accomplished with inadequate attention to consequences. Neglecting them was possible until they grew to become threats. To avoid them in the future and create a world in which a balance is achieved between the welfare of society and the welfare of the (planetary) ecosystem of which humans are a part requires vision of a desirable future (Steffen et al. 2011). The Future We Want conference demonstrated that such a vision is emerging across the planet. Many examples are given in this chapter of actions taken to develop water resources to meet human needs while ensuring the continuation of the essential services provided by our ecosystems. The future is becoming brighter.

4.7 Summing It Up

The environment, planetary or more local, is a system of ecosystems, each of which is a collection of interacting, interdependent components. Humans are one of those components. We depend on our ecosystems for the services they provide and for the quality of our lives. Yet through our increasing numbers and per capita consumption of goods and services and the natural resources that provide them, and the wastes that consumption generates, we are stressing our sensitive environment. Together with our deliberate modifications of our environment, for good or evil, this is putting our ecosystems out of balance and sometimes changing them irreversibly. All humans are affected by the consequences, but the water-scarce poor suffer the most and are the least able to cope. Humans alone among the species on this planet have the capacity to create a new and healthy equilibrium through observation and reflection. This will require:

- An appreciation of the value and the sensitivity of the ecosystems of which we are a part
- Continuous monitoring of changes in the environment
- Agreement on a broad vision of the new desired environment
- Taking action through a collaborative effort involving levels of society to reach that broad vision

Those knowledgeable about the water cycle, and the management of water resources and the services they provide must inform and participate in this process.

Integrated Urban Water Resources Management

5

Most of the world's demographic increase will take place in urban areas. Some of the most complex challenges that humanity face, now and in the decades to come, are related to this rapid urbanization. Urban centers offer opportunities for social and economic development, but infrastructure is needed to make that happen. The provision of a safe drinking water supply, sanitation, and storm water drainage integrated with the other urban services, such as the transport and disposal of solid waste and housing development, will be fundamental to achieving and maintaining well-functioning, livable cities. At the same time, managing solid and liquid waste and renewing aging infrastructure will require substantial investments. New technologies should help lower the costs and allow a more inclusive access of the poor to safe drinking water and sanitation, especially where it is most needed in the slums of the megacities of the developing world.

5.1   Urban Centers and Urban Water Systems

The geographic and social distribution of wealth in any given country or region is generally uneven. This observation is valid regardless of the size of a country. At the top is the primary city and at the bottom is the rural area. Sometimes this situation occurs within large urban centers. Hence, urban centers have always attracted people and this tendency will continue in the future. However, both in developed and developing regions the distribution of this economic landscape is a continuum of densities that gives rise to a variety of places with the primary city leading the group. Below the leading city there is a spectrum of smaller cities, towns, and villages as shown in Fig. 5.1.

Urban centers can also be viewed as production, consumption, and housing areas with high population densities, which are sustained by natural resources and products from remote areas normally located outside their political boundaries (Rees 2003). All cities depend on water, food, and energy inputs. As a direct consequence of the use of food, water, and energy, together with other resources that flow into the

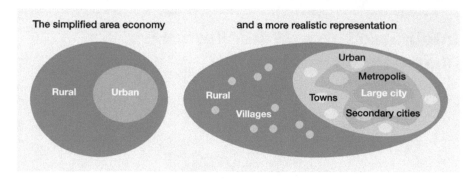

The simplified area economy   and a more realistic representation

Rural   Urban

Rural   Towns

Urban
Metropolis
Large city
Secondary cities

Villages

Fig. 5.1 Representation of urban centers and rural areas. *Source*: World Bank (2009)

city, there are solid, liquid, and gaseous residuals. There have been many attempts to define an urban center in environmental or ecological terms (Mcintyre et al. 2000). However, as observed by Odum (1997), urban centers can, from one perspective, be viewed as parasites of the countryside from which they obtain their inputs of food, water, energy, and other materials. They then degrade their environment by discharging the waste products produced from all their activities. The proper treatment and final disposal of these residuals is still an unresolved matter in many urban centers in developing regions. In particular, the lack of a safe drinking water supply and proper wastewater management poses additional public health problems. Yet cities are also centers of research and innovation, and, as engines, they can provide the rural areas with new knowledge and technologies. Equally significant, they provide markets for products that come from the rural areas and thereby provide income and opportunities for dwellers in rural areas.

Urban systems can be thought of as containing six interacting components—social, economic, environmental, land-use related, infrastructural, and institutional. Social and economic factors are the main drivers of urban centers since they create opportunities for jobs and better living conditions, but they also can cause non-desirable environmental consequences. The prosperity of nations is intimately linked to the "net" prosperity of their cities. No country has ever achieved sustained economic growth or rapid social development without urbanizing. Countries with the highest *per capita* income tend to be more urbanized, while low-income countries are the less urbanized. Thanks to an increase in productivity, urban-based enterprises normally contribute large shares to GDP (UN HABITAT 2011). For the future, city planners and the public at large must also cope with the downside, the non-desirable effects of wealth creating processes.

To meet the minimum requirements for being livable, the use of urban space should follow a land-use development plan that integrates the different structural and nonstructural aspects of urban centers in different areas. These aspects include housing, transportation, water (supply, treatment, and distribution systems), treatment of residuals, and storm water management. Urban land-use plans must allow for the mobility of, and communication among, urban dwellers.

The socio-environmental components contained in a land-use development plan are essential to allow humans to live harmoniously with the natural environment within the built urban ecosystem. Adequate institutions must be in place to guarantee that all actions foreseen in an urban land-use plan are properly implemented.

5.1.1   Current Conditions in Urban Centers

Today slightly over half of the world's population lives in urban areas. The UN expects that by 2050 almost all the world's demographic growth will be in urban areas. This is in contrast to the pattern that prevailed in the post World War II era, when there was a balanced split between urban and rural areas. By 2050, urban dwellers will account for between 85 and 90 % of the population of the more developed world and 65 % in the less developed regions. An average 7 out of 10 people will live in settlements referred to as urban (UN HABITAT 2011).

A salient feature of this rapid urbanization in the developing world is the growing income inequality within the city as well as between cities and rural areas. UN HABITAT finds that the population of slum dwellers around the world continues to grow at a rate of 10 % annually. Currently, about 30 % of the urban population in the developing regions lives in slums. UN HABITAT also finds that nationwide aggregates of Gini coefficients (Gini 1912), which measure the inequality of earnings between the rich and the poor, do not adequately reflect disparities in general urban or city-specific incomes. For instance, in the United States, where the 2005 national coefficient was 0.381, it exceeded 0.5 in many major metropolitan cities, such as Washington, DC, New York, NY, and Miami, FL. In developing regions, the inequality is even starker (e.g., 0.75 in Johannesburg). Again according to UN HABITAT (2011) the cities that are mostly poor have the least disparity. The implications of this growing pattern for urban water management are severe. How can the poor be provided with sustainable safe water supplies and sanitation services that they cannot afford?

The provision of adequate water supplies and sanitation services to this growing urban population, together with other urban services, such as the proper disposal of solid and liquid wastes, the provision of appropriate public transportation and housing for the poor, and the development and management of urban storm water drainage systems, are challenges faced by the majority of the urban centers in less developed regions. In those regions, the lack of proper wastewater treatment results in the direct discharge of industrial and domestic sewage into public spaces and eventually into recipient water bodies (Tsegaye et al. 2011). The consequences, in terms of public health, can be dramatic. Today two million children under the age of five die each year from water-related diseases, while the estimate of the global economic return on sanitation spending is USD 5.50 per US dollar invested (WHO 2002).

Urban drainage challenges are equally important. Today, the fragmentation of responsibilities and, especially, the lack of a coordinated responsibility among different sectors in the municipalities (solid waste management, drainage, transportation, and housing) often reduce the effectiveness of urban drainage development and management measures.

5.1.2 Aging Urban Infrastructure in Developed Regions

Urban water supply and sanitation systems in operation today in developed regions were constructed, in some cases, more than 100 years ago. These systems are facing important challenges especially from the point of view of public health. For example, the US National Research Council recently assessed and identified the risks associated with water distribution systems in the United States (NRC 2006). One of the most important issues facing distribution systems today is the problem of renewing old water and sanitation network systems. Old systems that leak face the risk of spreading waterborne diseases. Most of the reported outbreaks associated with distribution systems have involved contamination from cross-connections and back-siphonage.

According to Means (2012), leaking pipes lose an estimated 30 billion liter/day in the United States. Such a huge loss has induced the US Government to change its policy from new construction to the rehabilitation/renewal of piping components to maintain service (ASCE 2009). The latest American Society of Civil Engineers' infrastructure condition report gave drinking water infrastructure a D score and concluded that upgrading drinking water systems in the United States will cost at least USD 11 billion. Interestingly, the general public most affected by failures in this infrastructure does not support any fee or tax measures that would raise the money required to pay for this work. They would rather pay more to buy bottled water! Perhaps when there is not enough water for flushing their toilets they might think about it some more.

5.1.3 Current Urban Water Management

From different points of view, urban water management should include access to safe drinking water and sanitation for all inhabitants, storm water drainage with mitigation of the negative effects of flooding and its relationship to housing and transportation, and the control of waterborne diseases and epidemics. As such, the proper management of urban water requires the integration of just about all activities taking place in the urban space.

Water supply, sanitation, wastewater treatment, storm water drainage, and solid and liquid waste management have been planned, delivered, and managed largely as isolated services. A variety of authorities, each guided by distinct policies and legislation, continues to oversee water subsectors at the city level. For example, the traditional urban water management model has failed to distinguish between different uses requiring different water qualities (Bahri 2012). As a result, high quality water has been diverted to indiscriminate urban water needs, including for fire protection and wastewater transport in sewers (Van der Steen 2006). This issue is not confined to city boundaries. Basin level management often neglects to acknowledge the cross-scale interdependencies in freshwater, wastewater, flood control, and storm water (Tucci et al. 2010).

A growing city in developing regions often does not have the capacity to supply all the water required. In such cases, the population finds its own solutions. One is by pumping water from ground water aquifers. This can create risks of pollutant contamination (when pumping from shallow aquifers) or excess salinity (when pumping in coastal areas). Another is by extracting water from upstream sources and/or water made available with the help of interbasin transfer schemes and delivered to urban areas, where it is used and polluted, then rechanneled downstream, often untreated.

Since, in most cases, urbanization proceeds from downstream to upstream in river basins or coastal areas, the water supplied from upstream sources or from ground water (or a combination of these options) should be from protected sites. But urban growth in the upstream direction can compromise the integrity of such sites. The analysis of surface and ground water supplies is an important task that is often not considered in water supply projects in developing countries. Sometimes, costs can be reduced by up to 40 or 50 % when an integrated analysis is considered (ANA 2010). Water that is used by the population and then discharged into streams or septic tanks may overspill to drainage systems and rivers. This can introduce a large pollutant load to the rivers and ground water. As a consequence, water in the polluted river or groundwater aquifer cannot be used safely as a source of water supply. Since typical urban development spreads upstream, most of the upstream basin may become polluted and the source of clean water lost. In addition, urbanization competes with agriculture for space and water. Urban centers tend to develop in the more or less sedimentary flat areas well suited for agricultural activities. At the time when cities were concerned with the possibility of attack from invaders, they were located inside walled areas on hills or mountains, leaving the fertile floodplains for food production. Today this has changed.

Water issues often remain disconnected from broader urban and regional planning processes. This is especially true in the rapidly growing, high-density, urban centers of the developing world (Angel et al. 2011) where the larger number of new arrivals comprises the poor and where urban sprawl becomes unlivable slums. Where alternative approaches to address these problems are available, city authorities and municipal managers frequently lack the institutional capacity required to implement them. The result is informal settlements and peri-urban sprawl (Bahri 2012).

The prevailing water supply and sanitation systems in operation today in developed cities are the same as those initiated by the Roman Empire. These systems normally abstract water from remote areas where the water is usually sufficiently clean that treatment to remove harmful bacteria is all that is required. This water is then transported over long distances to the urban centers where it is treated to potable water standards. This high quality water is then distributed to all users, including those who do not require such high quality. The effluents that result from the use of this water are then discharged directly into the environment, or they are treated again to meet effluent quality standards and then discharged back to the aquatic environment.

In the early 1900s, at a time when the world population was less than two billion and predominantly rural, this approach successfully protected people from enteric diseases and flooding. The accelerated growth of population in the last century, and the associated urbanization of the planet, has shown that this approach is no longer working in less developed regions. Urbanization increases impervious areas and channelization, which, in turn, increase the flood peak and the flood frequency for the same rainfall. Urbanization also increases the flow velocity, solids production (sediments and solid waste), and liquid waste. Because of the lack of technical and institutional arrangements, most of the waste flows to the rivers, increasing their flows and decreasing their quality, since most of the storm water runoff carries pollutants from the land over which it flows. Pumping ground water, together with the reduction in infiltration because of the more impervious areas, could create subsidence in lowland areas. This decreases the drainage capacity by gravity and thus increases the frequency of flooding. Urban areas can be flooded by intense storms, flows from upstream areas, and by the sea (in coastal zones).

In summary, urban water management involves regulating both quantity and quality to meet desired standards. Having enough water to drink is not sufficient. Its quality also has to be acceptable. Contamination of water supply sources (streams and groundwater) typically results from the direct discharge of untreated effluents and diffuse pollution loads. Deterioration of water quality, resulting from a lack of sewage treatment, has created potential risks to the health of urban populations. A critical issue has been the occupation and contamination of source areas for urban water supplies (Tucci 2009).

Pollution control involves reducing, when physically and economically feasible, the production of waste streams, the collection and transport of pollutants, and, eventually, their treatment. The sanitation divide in cities—where some enjoy the benefits of pollution control and others do not—introduces not only inequities but also a human health hazard. Where traditional pollution control measures are not available it may be possible to consider implementing more local solutions. Technology is providing an increasing number of options for home sanitary arrangements that make it possible to handle domestic liquid and solid wastes onsite and to transport the residue to safe disposal sites. This technology includes dry sanitation in which the separation of urine and feces will allow the treatment of the wastes onsite. However, this technology may be difficult to implement in the existing large urban centers of the developing world. Given the sensitive cultural aspects, it is vital that sanitary solutions are planned and implemented with due consideration to social acceptance.

Partial solutions can be ineffective. The use of flush toilets—that are important for hygiene and sanitation in houses—may discharge into an open drain and this, in turn, can create even more unhygienic conditions and be the cause of disease in the neighborhood.

5.1.4   Urban Centers of the Future

Forecasting the future of urban centers and how that future is affected by water management policies is very difficult to do, as are attempts to make socioeconomic or political forecasts. Still, based on trends, there is a high probability that urban centers will increase in number and in population in the future. This means that more and more people will be living in urban areas for the simple reason that cities offer improved professional, educational, social, and cultural opportunities.

A complicating factor, especially for developing regions, is the impact of climate variability on the livability of urban centers in the future. Typically 30–50 % of the population in these urban cities lives in extremely poor housing and neighborhood conditions in degraded areas subjected to flooding and landslides. Long term investments are needed to minimize the problems that they will face in the future. Climate change is expected to increase such vulnerability in many cities around the world, as it is expected to increase the incidence and the intensity of hydro-meteorological extremes, such as windstorms, droughts, and floods (UNDESA 2009).

Although there are complicating factors for cities in the developing world in the future, there are also important opportunities. The United Nations Population Fund (UNFPA 2007) reports that most of the rapid urbanization in cities in Africa will take place in emerging towns and villages and not in the megacities. For every large town (50,000–200,000 inhabitants) there are ten small to medium-size towns (2,000–50,000 inhabitants) and they are expected to grow at least fourfold in the next 30 years (Pilgrim et al. 2007). Therefore, there are opportunities to address future urban water supply problems by involving water associations, autonomous town water boards, and small-scale private water companies in urban planning. Certainly an important component related to the solution of this problem is training and capacity building for the proper operation and maintenance of these small utilities.

5.2   The Need for Integrated Urban Water Resources Management

Integrated water resources management has been discussed for at least 30 years. Beginning immediately after the creation of the International Water Resources Association, this approach has been pursued by planners and analysts throughout the world with some degree of success. Today, the complexity of water resources planning has posed new challenges to this concept. Some attempts to define more precisely this concept are found in the literature (GWP 2000). The difficulty in implementing a truly integrated approach to water resources planning and management is its multidisciplinary nature.

There is no doubt that the management of water resources will play a significant role in urban expansion, and in the economic development and the quality of the natural environment of the regions where cities are located. Water policy and management must include an increasing inter-sector competition for water, a need to cater to multiple rather than single sector requirements, and a strategy that pays attention to water "from rain to drain"—from the sites where precipitation falls to the downstream sites where drainage from an accumulated number of uses occurs.

How water resources are managed affects the social, economic, political, and natural environments. Today, each of these aspects of our overall environment is considered important. In the past, however, water resources were often developed to serve a single purpose, such as hydropower, irrigation, navigation, or water supply. Large single purpose projects, developed in North America and Europe, led to an almost complete use of the available sites for dams and other related civil works, such as canals, dikes, interbasin transfer schemes, and the like. This was the common practice until the late 1960s when pressure from nongovernmental organizations (NGOs) and the public resulted in national environmental laws that forced natural resource planners and managers to consider environmental impacts along with economic ones. Today, environmental awareness is widespread in almost every country, developed and developing, throughout the world.

This multipurpose, multi-objective focus has motivated the establishment of new decision-making mechanisms to help planners and managers identify efficient trad-eoffs among the values of multiple objectives associated with multiple alternatives. The old, single cost-benefit analysis and top down approach are being gradually substituted by multi-objective, multi-decision-maker models in which stakeholders, including NGOs and government agencies, all participate democratically in the decision-making process.

Examples of this new decision-making process are the river basin committees that work as "water parliaments". This new paradigm requires a restructuring of existing institutions and the concourse of professionals of different backgrounds to handle the difficult task of conducting planning and management initiatives in an integrated fashion. Traditional engineering and economics professionals will have to work in close collaboration with social scientists and politicians.

The urban environment requires the use of this new paradigm in the most urgent way. Megacities in developing regions are under social as well as environmental stress. These megacities are under financial stress to renew their infrastructure, including that for water supply, urban drainage, and sanitation. These issues must be duly analyzed and innovative proposals for solving water-related problems in coordination with other urban service needs must be in place as soon as possible and certainly by 2050. Addressing the complexity of urban systems will require increasingly sophisticated knowledge of solutions that have both technical and social dimensions. Coordination is a key requirement of an integrated approach.

Given the complexity of urban systems, it is appropriate to question just how large an urban system should be to maximize the likelihood of achieving useful results from any systems analysis. Much will depend on the multiple institutions involved in the analysis and the scopes of their authorities and responsibilities as well as their information needs. While many favor the theoretical advantages of comprehensiveness, integration, and unified planning and management, successful practices seem to lag behind. It is important to recall that an on-going process of "complexification" (Vlachos and Braga 2011) creates a great diversity of interests, differences in resource accessibility, complex organizational arrangements, and competing and conflicting water policies. Finally, the shifting, over time, of goals, preferences, priorities, political consensus, leadership, and sponsors adds to the challenges facing urban systems modelers and planners.

The increase of urban centers will surely motivate the development of new approaches to implementing integrated urban water resources management (IUWRM). IUWRM is not a set of quick fixes for isolated urban water management problems. Rather, it reframes a city's relationships to water and other resources, and rethinks the ways in which they can be overseen (Bahri 2012).

According to Bhari (2012: 36), IUWRM is, in essence, a process that:

- *Encompasses* all the water sources in an urban catchment: blue water (surface water, ground water, transferred water, desalinated water), green water (rainwater), black, brown, yellow, and gray water (wastewater), reclaimed water, storm water, and virtual water
- *Matches* the quality of different sources (surface water, ground water, different types of wastewater, reclaimed water, and storm water) with the quality required for different uses
- *Considers* water storage, distribution, treatment, recycling, and disposal as a cycle instead of discrete activities, and plans infrastructure accordingly
- *Plans* for the protection, conservation, and exploitation of water resources at their source
- *Takes* into account the other, nonurban users of the same water resources
- *Recognizes* and seeks to align the range of formal (organizations, legislation, and policies) and informal (norms and conventions) institutions that govern water in and for cities
- *Seeks* to balance economic efficiency, social equity, and environmental sustainability

It is important to include in this very comprehensive definition the close relationship between land use and water use in the urban environment. This is critical, for instance, when dealing with urban flooding as will be discussed later in this chapter. Besides, it is also important to consider the integration of water management with other urban public services, such as transportation, energy, and solid waste collection and disposal. The proper management of urban flooding needs to consider these elements all together from a technical, as well as an institutional and political-institutional point of view. A summary of the required changes from traditional urban water management practices to the integrated urban water management approach is given in Table 5.1.

5.2.1   Water Supply and Sanitation

The new IUWRM will require that water supply and sanitation systems and the use of energy and chemicals be more efficient and make use of the nutrients that exist in the sewage. A series of new technological approaches are envisaged. These include capturing local water resources through techniques, such as rainwater harvesting and distributed storm water management, reducing net water use through water conservation, water reclamation, and recycling technologies, becoming energy neutral by both reducing energy use through distributed urban water management and by extracting energy and heat from the wastewater stream, and recovering nutrients (Daigger and Crawford 2007).

Experience in water-short areas indicates that application of these technologies can reduce domestic and commercial indoor water use substantially—from over 400 liter *per capita* per day (lpcd) to between 120 and 150 lpcd. Reduced water use also means significantly reduced used-water volumes, which can

Table 5.1 Comparison of past urban water management and integrated urban water management (IUWRM)

Past urban water management	Future IUWRM
Water and waste water systems are based on historical rainfall records	Water and wastewater systems rely on a multiple sources of data and techniques that accommodate greater degrees of uncertainty and variability
Water follows one path from supply to single use, to treatment, and disposal	Water can be reclaimed and reused multiple times, cascading from higher to lower quality
Storm water is a nuisance to be conveyed quickly from urban areas	Storm water is a resource to be harvested as a water supply and infiltrated or retained to support aquifers, waterways, and vegetation
Human waste is a nuisance to be treated and disposed of	Human waste is a resource to be captured, processed, and used as fertilizer
Linear approaches deploy discrete systems to collect, treat, use, and get rid of water	Restorative and regenerative approaches offer integrated systems to provide water, energy, and resource recovery linked with land-use design, regulation, and community health
Demand equals quantity. Infrastructure is determined by the amount of water required or produced by the end users. All supply-side water is treated to potable standards; all wastewater is collected for treatment	Demand is multifaceted. Infrastructure matches the characteristics of the water required or produced for end users in sufficient quantity, quality, and level of reliability
Gray infrastructure is made of concrete, metal, or plastic	Green infrastructure includes soil and vegetation as well as concrete, metal, and plastic
Bigger is better; collection system and treatment plant are centralized	Small is possible; collection systems and treatment plants may be decentralized
Standard solutions limit complexity; water infrastructure consists of hard system technologies developed by urban water professionals	Solutions may be diverse and flexible; management strategies and technologies combine "hard" and "soft" systems devised by a broad range of experts
Utilities track costs alone and focus on accounting	Utilities evaluate the full array of benefits from investment and technology choice and focus on value creation
The standard is a business-as-usual toolkit	An expanded toolkit of options includes high-tech, low-tech, and natural systems
Institutions and regulations block innovation	Institutions and regulations encourage innovation
Water supply, wastewater, and storm water systems are physically distinct. Institutional integration occurs by historical accident	Water supply, wastewater, and storm water systems are intentionally linked. Physical and institutional integration is sustained through coordinated management
Collaboration equals public relations. Other agencies and the public become involved only when approval of a predetermined solution is required	Collaboration equals engagement. Other agencies and the public are actively involved in the search for effective solutions

Source: Bahri (2012)

facilitate added innovations as will be discussed below. Reduced water consumption means that local water sources can provide a higher proportion of the urban water supply (Daigger 2012).

An important question related to the provision of water supply and sanitation services is related to the spatial nature of the network and related treatment facilities. This is not an issue related to decentralization or centralization. To decentralize or to not decentralize is not the question. Economies of scale normally will point to larger systems. However, it is difficult to define precisely the boundaries when decentralized solutions are more cost-benefit effective or when other arguments should be taken into account to make a decision (Starkl et al. 2012). This question becomes more relevant in developing regions where some suggest that the linear approach of the more developed nations is too capital intensive and should not be followed.

As rightly pointed out by Starkl et al. (2012), one of the main constraints to the implementation of decentralized systems is the operational challenge. While centralized systems are, in general, managed by institutions with strong technical and managerial capabilities, decentralized system can lead to poor maintenance, which will ultimately affect the end user.

In less densely populated areas, decentralized systems may be considered more effective, but in most urban situations a combination of both centralized and decentralized approaches can be a viable option. The question whether decentralized or centralized solutions have a better potential for the proper management and/or recycling of waste byproducts depends, in part, on governance and socioeconomic conditions, users' perceptions and participation, and regulatory enforcement (Starkl et al. 2012).

5.2.2   The Role of Technology

Recycling, reclamation, and reuse. Water reclamation is an important source of water in water-scarce areas and its reuse can contribute significantly to meeting urban water needs (Daigger 2007). Used water can be treated to meet any quality requirement, including for potable use or for industrial use, both requiring ultra-pure water. Advanced treatment technologies are increasingly becoming realistic choices for water, wastewater, and storm water treatment. They help cope with stringent standards, enhance capacities, and address contaminants that cannot be managed with conventional technologies (Bahri 2012). An evolving approach is source separation, which recognizes that different water qualities are appropriate for different uses. This general concept has been used for several decades with dual distribution systems, where reclaimed water is distributed for urban agriculture with high value crops and is now being extended, recognizing the small proportion of water that actually is needed to meet potable water standards—of the order of 30–40 lpcd.

Given their better capabilities and performances, membrane-based technologies and membrane bioreactors are penetrating the markets in many water-scarce regions because they enable the recycling of wastes and the use of alternative sources (such as brackish water and seawater). Recent studies (DeCarolis et al. 2007) suggest that a steady decrease in membrane costs has occurred during the last decade. Robust and durable membrane materials, as well as low-energy membrane systems (in some cases gravity driven) are being developed. Other technologies, such as photovoltaic systems with a renewable power source (solar driven) and oxidation processes, which can be enhanced with catalytic processes in combination with membrane systems, are coming onto the market. This trend will enable utilities to upgrade their systems (Bahri 2012).

Nanotechnology concepts are being investigated for higher performing membranes with less fouling properties, improved hydraulic conductivity, and more selective rejection/transport characteristics. Microbial fuel cells are emerging as a potential breakthrough technology that will be able to capture electrical energy directly from the organic matter present in the waste stream. Although these technologies are still in the early stages of development, and significant advances in process efficiency, demonstration, and production to commercial scale are necessary, they have the potential to enhance treatment-process performances and improve the efficiency of resources use (Daigger 2008).

Natural treatment systems. Natural treatment systems (NTSs) use natural processes to improve water quality, maintain the natural environment, and recharge depleted groundwater sources. These systems are, in general, cost-effective. Wetlands and oxidation ponds are examples of NTSs that are increasingly being used to treat and retain storm water, wastewater, and drinking water flows. NTSs have the advantage of being able to remove a wide variety of contaminants at the same time, which makes them a total treatment system on their own, and they are increasingly being used for water reclamation.

Source separation of waste streams. Key to the application of most of the new treatment technologies is the separation of the different flows of wastewater according to their pollution load. Most of the contaminants of concern in wastewater are contained in black water. For example, most of the organic and microbial contaminants are generated from fecal matter (which accounts for only 25 % of domestic waste), while most of the nitrogen and the emerging contaminants, such as pharmaceutically active compounds and endocrine disrupting compounds, are present mainly in urine.

New technologies, such as vacuum sewage systems and urine separation toilets, which reduce most of the nitrogen and trace organic contaminants, have made it possible to handle a small and concentrated amount waste.

Desalination. Desalination of brackish water and seawater is becoming increasingly economical, thanks to advanced membrane technologies and improved energy efficiency (Bergkamp and Sadoff 2008). The cost of producing desalinated water is now estimated at USD 0.60–USD0.80 per m^3 (Yuan and Tolb 2004). In countries that have exhausted most of their renewable water resources, desalinated water meets both potable and industrial demands. However, its use in agriculture remains limited. Desalinated water is already being used for the cultivation of high value crops in greenhouses.

These technologies create opportunities for the reuse of gray water at the source and the recovery and reuse of nutrients. They also reduce the cost of extensive sewer systems and minimize, or may even avoid, the use of clean

Table 5.2 Innovative technologies and its benefits for integrated urban water management (IUWRM)

Innovative technology	Benefits for IUWRM
Natural treatment systems	• Adds multifunctionality (integrated treatment and environmental functions)
	• Improves environmental quality
	• Utilizes natural elements, features and processes (soil vegetation, microorganisms, water courses, etc.)
	• Is robust and flexible/adaptive
	• Minimizes the use of chemicals and energy
	• Promotes water reuse and nutrient recovery
Nanotechnology and microbial fuel cells	• Provides access to a cheap "green" energy source (enables the capture of electrical energy directly from organic matter present in the waste stream)
Membrane bioreactors (wastewater)	• Enhances new strategy for water management to move toward water use
	• Reduces plant footprint
	• Can easily retrofit wastewater treatment processes for enhanced performance
	• Offers operational flexibility (amenable to remote operation)
	• Manages environmental issues (visual amenity, noise, and odor)
Membrane technologies (both water and wastewater)	• Promotes decentralized systems which minimize the environmental footprint
	• Enhances contaminant removal and encourages water recycling
	• Minimizes the use of chemicals
	• Improves system flexibility and permits small-scale treatment systems
Source separation	• Promotes water reuse and nutrient recovery
	• Promotes small (decentralized) systems that can be easily managed
	• Avoids the complications and cost of dealing with mixed wastes
Anaerobic fermentation (UASB)	• Produces biogas
	• Promotes the recovery of energy from wastewater

Source: Bahri (2012)

water to carry waste. An overview of innovative technologies that support IUWRM is provided in Table 5.2.

Demand management. One way to cope with a lack of water availability in cities without capital to invest in water infrastructure is to manage the demand for water. Urban dwellers can become more efficient in their use of water for different activities. Unfortunately, some examples from developing countries, for example, in New Delhi, show that urban *per capita* use is coming closer to that of the richer parts of the world, such as Australia. It is unfortunate, but the use of water grows with wealth. It is the price of becoming affluent—households become huge water guzzlers, and have no inkling of, or are not concerned about, the huge consequences of their newfound water use. When such behavior becomes a part of people's lifestyles, urban demand for water can actually compete with the consumption of water for other water uses, particularly in agriculture.

Australian cities' *per capita* domestic water use, at 320 lpcd, is second in the world after the United States. But now Australia, grappling with severe drought-like conditions, has taken the lead through its Water Efficiency Labelling and Standards scheme (WELS) and legislation. The United States has a similar scheme—WaterSense—promoted by the Environmental Protection Agency. But this is not mandatory; rather, it promotes consumer choice. In contrast, water-stressed Singapore has introduced its own version of WELS, which, like the Australian program, applies to all kinds of sanitary water-using appliances in households—from showers to toilets (see Box 5.1).

Box 5.1 Urban Water Conservation Programs in California

In industrialized nations, where extensive water infrastructure has been built, there has been little attention paid until recently to the efficiency of water use. The high reliability and low cost of much municipal water use has encouraged inefficient technologies and uses and a lack of awareness of the water constraints more common in water-scarce regions. This is changing. As populations have increased, and as pressures on limited water resources have grown, even regions formerly considered water rich are expanding efforts to reduce waste and improve water use productivity. As a result, urban water use efficiency has been increasing in many regions, leading to a leveling off, or even a decline, in both total and *per capita* water use. In northern California, water demand for residential, industrial, and commercial users from the East Bay Municipal Utilities District, which provides water for nearly 1.4 million residents, has been flat for over 2 decades despite continued population growth. Thus total *per capita* water use has dropped significantly statewide. From 1995 to 2010, the average *per capita* urban use of water has fallen by nearly 25 %. Despite this progress, there is still considerable room for improvement. The Pacific Institute estimated that the potential for cost-effective urban water savings is still around 30 % of total urban water use, and such estimates have been incorporated into official state water plans (Gleick et al. 2003; CALFED 2006; California Department of Water Resources 2009). These reductions can be achieved through improvements in the efficiency of indoor plumbing fixtures and appliances, improvements in industrial processes, and reductions in the use of water in outdoor landscaping, which accounts for around half of the statewide urban water use. The savings achieved in California have also been mirrored in other parts of the world. *Per capita* residential water use in the early 2000s in Australia, for example, after suffering from extreme drought, dropped to less than

(continued)

Box 5.1 (continued)

265 lpcd in Queensland, and was only slightly higher in other parts of the country. In Israel and Spain, urban water use was as low as 320 lpcd and 290 lpcd respectively.

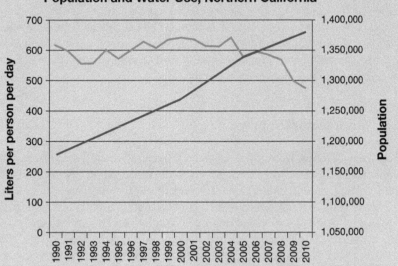

Population and Water Use, Northern California

Population (red) and *per capita* urban water use (blue) for the East Bay Municipal Utilities District in northern California. Water use has been level or declining, despite population growth. This water includes all water for residential, commercial, and industrial uses, including major refineries and other significant industrial uses. Residential use is typically around half of this total.

Source: http://www.usbr.gov/mp/watershare/wcplans/2012/EBMUD_2011_WMP2.pdf

Household level water audits suggest changing trends with growing affluence. In the UK, the Department for Environment, Food and Rural Affairs has found that an average household uses 55 % more water today, than they did in 1980. Manufacturers are responding to growing consumer demand and, more importantly, the growing stress on water. There is a range of fixtures—from dual flush to waterless urinals and taps for green building designers. But these water devices are very expensive. And in the situation where water is still poorly priced, there is little or no incentive for anyone to invest in higher capital costs.

A related problem for efficiency is the engineering specifications for the pipes that carry sewage from the homes. It is argued that current sewer lines are designed for a certain flow and, therefore, need the excess water to carry away the human excreta through the pipes. Reducing the water inflow will clog the drainage system. This issue was raised in Australia, where it was also suggested that, in the current design, flushing not only clears the refuse from the pan but also transports the waste through the drain to the main sewer lines. But Australian engineers found that this was not the case. Blockages in the main sewage lines had much more to do with the state of disrepair of the system than with the water inflow. But clearly this flow blockage problem will not exist in areas where sewage lines are nonexistent, as is the case in many parts of the country. It also provides an opportunity to revise standards for sewerage and plumbing so that new city systems can be built for the future, a future with less and less water, but more and more wealth.

Another important aspect related to demand management is public awareness and the use of economic instruments to incentivize more conscious water consumption. This may include differentiated tariffs for consumption levels or benefits for those reducing their consumption over time.

5.2.3 Stormwater Drainage

The drainage of storm water from urban areas has traditionally been accomplished by constructing storm sewers, through which the urban runoff is conveyed directly to the receiving water. Urban drainage before the 1970s was designed to transport storm water away as quickly as possible. However, during the 1970s more attention was drawn to the quality of urban runoff. The pollution content in storm water and its impact on receiving waters became a major concern. Measures were taken to protect receiving waters from polluted urban runoff. In the 1990s the concept of sustainable development was introduced. In this concept the social dimension of urban drainage came into focus (Stahre 2008).

The transition from traditional to sustainable urban drainage can be viewed as a sequence of steps:
- From the consideration of only quantity prior to 1975
- To both quantity and quality through 1995
- The addition of amenity values from 2005 to the present day
- Present day inclusion of integrated water management as well

What was "traditional urban drainage" prior to 1975 has now become "sustainable water management" (Stahre 2008).

The characteristic feature of sustainable urban drainage is that the quantity and quality aspects of the runoff are handled together and included with the social aspects of the drainage. For instance, many drainage approaches can be incorporated into the urban form to provide aesthetic and artistic amenities, such as landscaped storm water runoff storage and treatment features (e.g., eco/green roofs, and rain gardens).

Urban floods may be the result of two different types of phenomena—local floods from the rainfall precipitated over the urban center that result in streamflows exceeding the network design capacity, and regional floods that occur because of the overflow of larger rivers flowing through the urban center. Local floods may also be increased as a result of poor maintenance of the storm flow network or because of the decrease in pervious land surfaces as shown in Fig. 5.2. Urbanization replaces

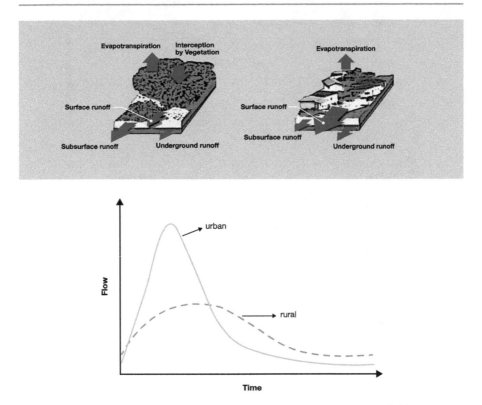

Fig. 5.2 The effects of urbanization in urban flood hydrographs. *Source*: Tucci (2009)

natural landscapes with impervious surfaces, such as roofs, walkways, and streets. As a result, water that would have infiltrated before urbanization now flows over impervious surfaces more rapidly until gutters, pipes, galleries, and channels are reached. The overland flow volume has increased, on average, from about 5–15 % of the rainfall to more than 60 % as shown in Fig. 5.3. Peak flow increases from three to seven times the discharge from natural conditions before urbanization took place (Yoshimoto and Suetsugi 1990).

In the case of local floods, the common procedure or coping with flood consequences has been to transfer the problem downstream by the channeling of small creeks and the construction of ditches and larger canals to make the flow move faster. This approach has the advantage of being less expensive, but has the disadvantage of transferring the problem of protecting upstream population to the downstream one.

Regional flooding affects the population living in the floodplains. Solving the regional flooding problem has been more difficult because of the associated costs of the necessary hydraulic infrastructure (canals, dikes, etc.) to make flows move more quickly to the downstream areas. Hence, in both cases, new approaches are necessary to resolve the urban flooding problem.

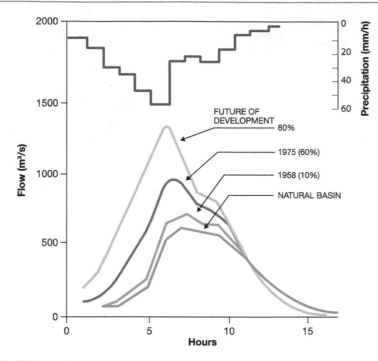

Fig. 5.3 Different flood hydrographs for the same rainfall intensity and different levels of urbanization. P=precipitation. *Source*: Yoshimoto and Suetsugi (1990)

The pioneering work of Gilbert F. White (1945) inspired a new policy of flood management in the United States and has influenced many states in that country to establish public policies for urban storm water drainage planning. The city of Denver, e.g., has initiated this type of approach after the catastrophic flooding events of 1965. In the majority of the cities in the State of Colorado, the institutions working in this area still take their actions based on these plans. In Europe, actions against floods go back centuries, as in the case of the Netherlands. After the 1980s it is possible to identify several experiences of urban flood control in the orient. From the 1990s some successful experiments have been developed in less developed countries of South America to minimize the impacts of landslides resulting from intense rainfall in the tropical environment.

The urban storm water drainage system is part of the existing urban public services, such as water supply and sanitation networks, electric energy and communication networks, public lighting, street cleaning, and parks and recreation areas. Hence, the urban community expects that its urban space be planned in an integrated manner. This integrated urban water master plan should take into consideration any sector plan for these different services. When the storm water drainage system is not considered in the general urban master plan at its outset, chances are that when this system is designed it will incur higher costs and will be less efficient. Relative to the other urban public services, the storm water drainage system has the peculiarity that storm water resulting from intense rainfall will flow regardless of the existence of an adequate and efficient drainage network.

These storm water drainage systems should take into consideration not only the structural measures (canals, galleries, and the like) but also the nonstructural measures, such as floodplain zoning, land-use control, and early warning systems. Furthermore these plans normally are based on scenarios of the expected hydrologic and economic development conditions of the area under study. The planning of these drainage systems should give priority to the retention of storm water in the upstream areas of the watersheds. This can be done by the construction of online and offline small reservoirs or linear parks, which can be periodically flooded. These structures, in large numbers and distributed all over the river basin, serve the purpose of retaining storm flows which would otherwise reach the main river courses in the basin causing extensive damage. The use of open channels in place of pipes and galleries to minimize costs are particularly important for developing countries.

It is clear that storm water drainage plans must be included in the integrated urban master plan. As a result, multidisciplinary teams of hydrologists, engineers, economists, geographers, and sociologists are essential in the development of these plans.

Storm water management can mitigate intense rainfall events and enhance local water sources. Cities that suffer from flooding have several options for urban storm water management. These include using retention ponds, permeable areas, infiltration trenches, and natural systems to slow the water down. Lodz, Poland, and Belo Horizonte, Brazil, both use such systems, and Birmingham, England, is experimenting with green roofs to achieve the same effect (SWITCH 2011).

Rainwater harvesting can help address water scarcity at the household level and may be easy and cost-effective to implement. Roof-water harvesting provides a direct water supply and can recharge groundwater, while reducing flooding. Such measures may be immediate solutions to accompany long term infrastructure improvements in water supply and drainage. To date, comprehensive documentation of the design criteria, costs, benefits, impacts, and constraints of large scale adoption is generally lacking and would be needed to evaluate the viability of scaling up (Bahri 2012).

5.2.4   Water and Energy Nexus

As mentioned in Chap. 7, distribution system pumping can account for more than 90 % of the energy use for many water systems and, therefore, a significant portion of the greenhouse gas emissions generated (directly or indirectly) by water utilities. A focus on maintaining pump efficiency is important not only for economic reasons but also for environmental stewardship.

Water conservation will be driven from several perspectives that include:
- The need to release more water for the maintenance of the aquatic environment
- Pressure to reduce water consumption
- Pressure to reduce energy use
- Pressure to reduce costs and greenhouse gas emissions
- Price elasticity effects—as costs rise, demand will be pressed downward (Means 2012)

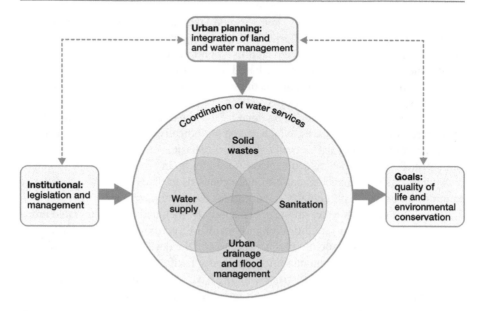

Fig. 5.4 Integrated urban water resources management. *Source*: Tucci (2009)

Consumers will demand more from both water and power utilities in the future, highlighting the interdependence between the sectors. Economies of scale will also play a definite role in the metropolitan areas and megacities of the future. The isolated utilities of today with their own billing systems, human resources and research standards, and efficiencies for water and energy will have to find ways of joining together to produce more efficient results.

5.2.5   Governance

Governance plays a very important role in making all the technical and managerial solutions proposed in urban plans work. It involves a collection of public institutions and public participation to create a management system, which guarantees coordination and cooperation among public, private, and NGOs. IUWRM requires the development of planning and management for all components of the urban water services (Fig. 5.4). These services are interconnected and require a high level of coordination and integration. Coordinating structures and forums will ensure communication between departments, between levels of government, and with communities and stakeholders.

Urban planners have an important role in helping governments overcome fragmentation in public policy formulation and decision-making. They can achieve this by linking planning with the activities of other policy sectors, such as infrastructure provision, and adopting collaborative approaches that involve all stakeholders in determining priorities, actions, and responsibilities (Fig. 5.5). This may involve new methods for interagency coordination and control of water use, such as a new institution or executive committees that have the authority and capacity to regulate and

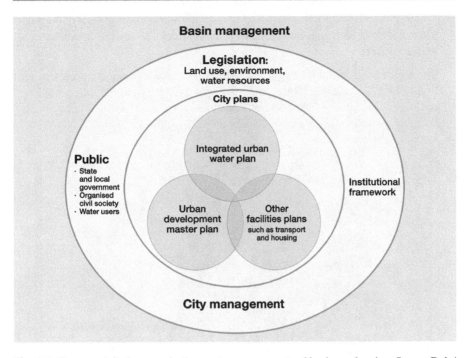

Fig. 5.5 Framework for integrated urban water management and land-use planning. *Source*: Bahri (2012)

enforce standards and procedures. Integrated urban water policies, based on participatory, democratic, and pluralistic governance, can secure sustainable development, particularly if governments adopt clear urban policies as an integral part of their economic policies (UNEP 2002).

Urban storm water management systems are based not only on plans, projects, and civil works but also on nonstructural measures that involve laws and regulations on construction, zoning, land-use parceling, as well as sanitary control and environmental conservation. These institutional measures are particularly valuable in low river sedimentary lands. These areas are typically flooded during certain periods of the year and their occupation must be restricted by defining land-use limits based on different degrees of flood risk, flood insurance systems, special tax incentives for not using these areas, and other non-structural measures.

This process needs to be participatory and transparent in order to gain the support of the community and local authorities. Normally, public participation is done through representatives of different organizations in the region, who should be able to discuss openly the alternatives. It is important to bring into the process all the interested parties, especially those poor or marginalized communities of the metropolitan regions (Box 5.2). As in this process there are different interests and positions being considered, multi-criteria decision-making models may help to move the political process toward consensus. Such methods can help decision-makers express their preferences in relation to any proposed alternatives and prioritize them according to preestablished criteria.

Box 5.2 Public Participation in the São Paulo Metropolitan Region Interbasin Water Transfer

São Paulo Metropolitan Region (SPMR) is the largest urban conurbation in South America and the largest industrial complex of Latin America. Its 18 million inhabitants, spread over an area of approximately 8,000 km², occupy 950 km² of urbanized area, which lies entirely within the Upper Tietê river basin, as shown below.

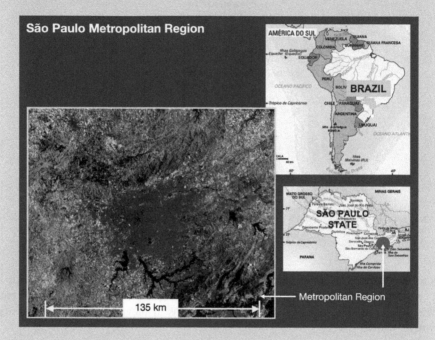

The disordered growth of the SPMR, notably from the 1950s onward, made the water supply system in the region inadequate, with successive situations of lack of water and rotations. In 1974 an interbasin transfer scheme from the neighboring Piracicaba river basin was concluded transferring up to 33 m³/s. This temporarily solved the water supply problem. At the time of this transfer no public participation existed in the decision-making processes in the water sector. With the economic development of the region since then, demands for opportunities to participate started to increase in both basins, the receiving upper Tietê and the donor Piracicaba. In 1997, with the passing of the Water Law (9433/97), a new water resources management system was implemented in the country. This system introduced the concept of planning and management on a river basin basis through river basin committees in which local governments, water users, NGOs, the academic community, and

(continued)

> **Box 5.2** (continued)
>
> professional associations participate in the decision-making process. In 2004, the authorization for this transfer expired and an intense negotiation process took place between the river basin committee of the Piracicaba River, the National Water Agency of Brazil, and the State of Sao Paulo Water Authority, DAEE. As a result, more water was apportioned to the Piracicaba basin and the Water Utility of Sao Paulo is now looking for new water sources to supply the growing demand of the MRSP. This example shows the importance of participatory processes in decision-making dealing with water issues.

5.3   The Way Ahead

The proper planning and management of urban waters probably will be one of the major challenges for humanity in the future. Demographers from the UN agencies all agree that future population growth will take place in the urban areas of the world with an expectation that by 2050, 7 out of 10 people will be urban dwellers. The provision of safe drinking water and sanitation for these areas will require new thinking about appropriate technology and financing. The cities of the less developed regions, where the highest population growth will take place, are where the need for innovative solutions to pressing water supply, sanitation, and storm water drainage requirements is greatest.

Demand management and appropriate maintenance of existing systems to reduce water loss and contamination should be pursued, both in developed and developing regions. In the latter, given the lack of financial resources, meeting these needs is especially challenging. The use of distributed systems to reduce capital costs has a potential in the future because of improvements in, and cost reductions of, novel water and sewage treatment systems. However, there is no general consensus on this issue and a detailed cost-benefit analysis needs to be done on a case-by-case basis.

IUWRM, as described, involves full consideration of water use and control of excess water in the urban planning process. This means that other public services, such as transportation, communication, public housing, and solid waste management, should consider water supply and sanitation as well as storm water drainage needs in an integrated way. This poses technological challenges, but technology is not the main obstacle to integration. Water governance plays a much more important role in putting IUWRM into practice. It involves creating the political will to allow full public participation in the decision-making process.

The challenge to achieve IUWRM in developing countries is much greater than in the more developed economies. Basically, the problem resides in the levels of education of the poor and more disadvantaged groups who are most affected, mainly by natural water disasters resulting from insufficient storm water drainage in their areas of residence. Hence, the professional community has to understand its role and participate more actively in the political processes related to urban issues. Through the participation of all involved stakeholders, the professionals, the government, nongovernmental agencies, and the public, IUWRM may actually become the rule in urban planning by 2050.

Water and Food Security: Growing Uncertainties and New Opportunities

6

The food challenge for humanity is to reliably produce increasing supplies and more varieties of food, mainly through higher yields using less water and other resource inputs in an environmentally friendly manner—and at reasonable and affordable prices for consumers. This challenge must be met as the climate changes, and as almost every input to the food production, processing and consumption chain increases in cost. This can only be done by mobilizing technology in all kinds of agriculture including smallholders, using all sources of water, and achieving much better coordination, cooperation, and partnerships among the major stakeholders involved: farmers, market operators, regulators, and consumers. Our way of life, well-being, and culture are intimately linked to how and where food is produced, what is produced, how we obtain it, how we prepare it, and how we eat it. The future of humanity depends on how food is—and will be—produced and provided.

6.1 Uncertainty Dominates

The world is rapidly changing. A new order is taking shape. Changes in socioeconomic and demographic dynamics in combination with technological and other changes are increasing the demand for a wide range of goods and services, all of which require more water and other resources. In the opposite direction, climate change and the associated global warming pose tangible challenges for societies. A predominant feature on the supply side of water management is more uncertainty and more pronounced amplitudes in freshwater resource availability, which adversely impact the reliability of goods and services that require freshwater. Food is one of those goods. Huge variations in water and food availability and growing competition over short and longer time-period objectives, as well as local and regional goals associated with water management and food production and use, will characterize the new world order.

Gulbenkian Think Tank on Water and the Future of Humanity, *Water and the Future of Humanity: Revisiting Water Security*, DOI 10.1007/978-3-319-01457-9_6,
© Calouste Gulbenkian Foundation 2014

Convergence and commitment to providing reliable and clean sources of water for all humans and the fair and efficient use of vital resources, including water, should be a prominent principle in a globalizing world. Geographically and geopolitically, a worrisome trend is that population growth, and its needs, will take place in areas where effective demand and often political bargaining power are weak. At the same time, water and other natural resources pose significant constraints for increased production and food security. The long-term achievement of food security is important for us and our descendants.

This chapter will discuss the major uncertainties pertaining to global food production and supply, and how those uncertainties affect the demand for food. As implicitly noted in the most commonly used definition of food security (FAO 1996), an active and healthy life presumes that the links between production and the actual intake of food are given due attention. Food security requires a systems perspective involving supply and demand. A focus on only one side of the equation, e.g., production, is not sufficient for proper understanding and effective policy. The chapter will also highlight opportunities in this regard and explore principles that can be translated into policies, strategies, and concrete action to overcome pending shortages.

6.2   Our Journey Toward Food Security

6.2.1   The Evolution of the Food Security Concept

The importance of securing enough food for people has been recognized since earliest times and in all cultures. In China, in the sixth century BC, the philosopher Lao-tzu wrote: "There is nothing more important than agriculture in governing people and serving Heaven." In addition, he admonished rulers who neglected the agricultural sector: "The imperial palaces are magnificent, while the farmlands are allowed to lie in waste and the granaries are empty. The governors are dressed elegantly, wearing sharp swords and eating luxurious food. They show off like robbers. How far away from the Tao!" (Norton 2004).

The alluvial plains of southern Mesopotamia provided the fertile soils that nourished the crops on which the world's first civilization was built. Agricultural surpluses freed laborers from the field, allowing them to become artisans or traders; the organization of essential irrigation projects provided a hierarchy of rulers and administrators; the export of grain paid for the import of luxury goods; and the subsequent rise of wealth attracted immigrants and merchants from the surrounding countryside. This basic condition of human existence was not lost on early economic theorists. Adam Smith perceived "as significant the relationship between productivity improvement in agriculture and the wealth of nations" (Johnson 1997). Indeed, the performance of agriculture over the centuries has made a fundamental contribution to present standards of living.

During the last 50 years, the continuing evolution of food security as an operational concept in public policy has reflected the wider recognition of the technical and policy issues involved. In fact, food security as a concept originated only in the mid-1970s. The initial focus, reflecting the global concern leading to the World Food Conference in 1974, was on the volume and stability of food supplies. Food security was then defined as "availability of adequate food supplies of food stuffs to sustain a steady expansion of food consumption and to offset fluctuations and prices" (FAO 2003).

The Green Revolution of the 1960s and 1970s provided the basis for a quantum leap forward in food production. It depended on applications of fertilizers, pesticides, and irrigation water to create conditions in which high-yielding modern crop varieties could thrive. Between 1970 and 1990, the amount of land under irrigation increased by one-third. The gains in production were dramatic: world grain yields jumped from 1.4 tonnes per hectare in the early 1960s to 2.7 tonnes per hectare in 1989–1991. The volume of world agricultural production doubled and world agricultural trade increased threefold. The continuous increase in yields over the last 50 years has outpaced population growth.

However, these technical successes of the Green Revolution did not automatically and rapidly lead to dramatic reductions in poverty and malnutrition. This observation led to the recognition that effective access to food is as critical as food availability. Thus, in 1983, the Food and Agriculture Organization of the United Nations (FAO) expanded its focus to include ensuring that "all people at all times have both physical and economic access to the basic food they need." This represented a shift away from maximizing crop yields to understanding the complex nature of hunger and famine.

The Nobel laureate Amartya Sen (Carlsson et al. 2009) pointed out that famine resulted not from lack of food in a given region but rather from political decisions that led to lack of "entitlements" by marginalized populations to gain access to the food they need. Such entitlements could consist of income, access to social protection, or smallholder production. Sen famously showed that food exports tend to increase from famine-affected areas, largely because poor people cannot afford to buy the food available. So increasing global food production and supply is not a panacea for avoiding famines. Naturally, a certain amount of food produced and supplied through various channels is a basic precondition for food security. The broader "freedom" to buy food, as influenced by local/national/international policies and market trends, was recognized as more important than household or even national production levels. Partly as an outcome of these new perspectives, during the 1990s the broader concept of livelihoods came to largely displace earlier attention on production and supply for food security. The struggles of poor people to survive in increasingly diverse and complex economies were recognized as a more appropriate starting point in preventing hunger than a narrow food-production perspective.

Until almost the mid-1980s most countries had striven to meet their food needs from domestic supplies. This quest for food self-sufficiency was justified primarily by (Ait Kadi 2000):

- The food crises and famines that raged in several parts of the world, particularly in Asia before the Green Revolution of the early 1970s.
- The hopes entertained by the Green Revolution. Thanks to increased production and productivity obtained, some countries have significantly reduced undernourishment, or even enjoyed food surplus.

- The number of people suffering from undernourishment was reduced from about 870 million people in the 1960s to about 770 million in 1995–1996 when it started to increase again.
- The risks inherent in dependence on international trade in agricultural products given the volatility of prices and protectionism.
- The Cold War and threats related to food embargoes. To gain their political independence countries were concerned about their food self-sufficiency as an issue of their sovereignty.

Typically countries with a surplus export only a fraction of domestic production. Indeed, the supply of agricultural commodities for the international market is small relative to total domestic production in exporting countries. Today seven billion people rely on 2.5 billion tonnes of grains and oilseeds. Roughly 300 million tonnes—about 12 % of total demand—enter into the world grain and oilseed trade. Another feature of agricultural trade is the existence of a limited number of exporting countries that dominate the international markets. Today, nine countries account for 90 % of the world's wheat exports, and just five countries account for 85 % of the world's exports of milled rice. In fact, together Thailand, India, and Vietnam produce 66 % of all milled rice exports. Such a concentration of suppliers exacerbates price changes on the world market, in the case of, e.g., abnormal weather conditions in these countries.

Even if trade in agricultural commodities is comparatively limited, the expansion of globalization and international agricultural trade in the 1980s led many countries to move from their food self-sufficiency objective and to rely more on the international market to cover strategic levels of their food needs. Ideally, the liberalization of agricultural trade widens the entire spectrum of economic possibilities—offering countries the potential to allocate their water resources more efficiently and to make the most of their comparative advantages. The challenge is to identify agricultural opportunities and act on them. However, appropriate trade policies require a clear understanding of the water and other natural resources preconditions, as well as the underlying economic competitiveness in any agricultural sector. In liberalized world markets, strong government policy is needed to encourage a sound and just development. Other essential ingredients are immense investments in infrastructure and technology, including marketing to increase productivity and improve quality in areas where production and productivity are currently below, or even much below, its potential.

Developing countries have transitional difficulties and economic and social adjustment costs. Agriculture plays an important role in their national economies. It provides employment and regional stability to a large portion of their population, but the need for technological and institutional reforms still exists. Globalization has exposed small farmers, in particular, to greater competition from international trade so forcing them to lower prices, even for their traditional crops. In Africa, e.g., small farms are being squeezed out of their traditional food crop markets in urban and coastal areas by cheaper imports, while being undercut in their traditional tree crop export markets by new competitors from Asia.

Small farmers are also increasingly being asked to compete in markets that demand much more in terms of quality and food safety; that increasingly come under the sway of supermarkets, processors, and large export traders; and that reflect far more international competition. Because small farms struggle to diversify into higher-value products, they must increasingly meet the requirements of these demanding markets, both at home and overseas. Barriers to market entry remain high. These barriers arise from geographic distance to national market centers, lack of market organizations and information, as well as the increased safety and quality standards of food processors and retailers. The regional and intercontinental integration of the agrifood system has been accompanied by a rise in the power and leverage of international corporations.

The 2007–2008 food crisis was triggered through a combination of various circumstances, including adverse weather conditions, price hikes on oil and oil products (e.g., fertilizers) and export embargoes on food, which resulted in increased costs of production and transport, and extreme variability in international food prices. Prices increased suddenly and significantly for nearly every agricultural commodity. For instance, at their peaks in the second quarter of 2008, world prices of wheat and maize were two times higher than at the beginning of 2003 and the price of rice three times higher. Uncoordinated and protectionist government responses led to enormous efficiency losses within the global food system, and hit the poorest countries and people the hardest. When food prices were particularly high, major producers imposed restrictions on agricultural commodities in order to minimize upward pressures on domestic prices. Although these export restrictions in more than 30 countries may have reduced the risk of domestic food shortage in the short term, they made the global market smaller and more volatile. Furthermore, these export bans provoked panic buying in the international markets that, in turn, deepened the food crisis. This distrust in markets has led many countries to reexamine the "merits" of self-sufficiency and to start rebuilding their national stocks (Headey and Fan 2010).

This food crisis swung the pendulum back toward increasing availability of food. Some countries began investing in agriculture to secure their food supplies. Large-scale acquisitions of farmland (also termed "land grab," which is also "water grab") in Africa, Latin America, Central Asia, and South East Asia have made headlines in a flurry of media reports across the world. For people in countries leasing out land, this new context may create opportunities for economic development and livelihood improvement in rural areas. But, in the absence of a code of conduct, it may also result in local people losing access to the resources on which they depend for their livelihood and food security (Daniel and Mittal 2009).

Today food security is recognized as a significant challenge, spanning a spectrum from the individual to the global level. As mentioned, the definition has gradually been broadened and also incorporates food safety and nutritional balance, reflecting concerns about food composition and minor nutrient requirements for an active and healthy life. Food preferences, socially or culturally determined, are considered— and local and regional traditions are important. But changes in food production conditions and increasing demand due to demographic and other socioeconomic shifts also imply that food security may be quite different from what can be achieved through local resources. The 1996 World Food Summit adopted a still more complex definition: "Food security, at the individual, household, national, regional and

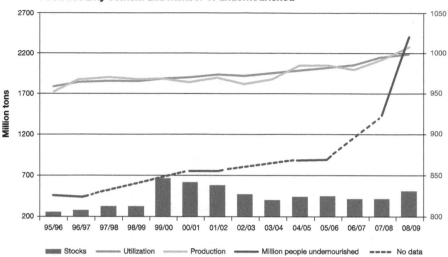

Fig. 6.1 An illustration of a parallel increase in number of people being undernourished and increases in food supply. 1995/1996–2008/2009. *Source*: Lundqvist (2010)

global levels [is achieved] when all people, at all times, have physical and economic access to sufficient, safe and nutritious food to meet their dietary needs and food preferences for an active and healthy life" (FAO 1996). This definition is again refined in The State of Food Insecurity (FAO 2001): "Food security is a situation that exists when all people, at all times, have physical, social and economic access to sufficient, safe and nutritious food that meets their dietary needs and food preferences for an active and healthy life." Therefore, food security should be considered in its four dimensions: first, availability; second, access; third, stability; and fourth, safe and healthy use.

The international community has accepted these increasingly broad statements of common goals and implied responsibilities. But its practical response has been to focus on narrower, simpler objectives around which to organize international and national public action, notably increased production. The declared primary objective in international development policy discourse is, however, increasingly on the reduction and elimination of hunger and poverty. Unfortunately, scant attention is paid to the challenge of how to improve access to the food that is actually available on the market for those who lack the means to buy what they need and want. Similarly, the low level of efficiency in the supply chain (see Chap. 2) is ignored.

The ambitious agenda of the first of the eight UN Millennium Development Goals is to cut the number of poor and hungry people in half by the year 2015. We are rapidly approaching this deadline but during the last decade or so, we have moved in the wrong direction in terms of reducing the number of undernourished. While hunger dropped between the 1960s and mid-1990s, a staggering number of people, approximately a billion people worldwide, face chronic hunger. As a comparison, an estimated 1.5 billion people (aged 20 years and above) are overweight and at risk of being affected by a range of bodily diseases (Fig. 6.1).

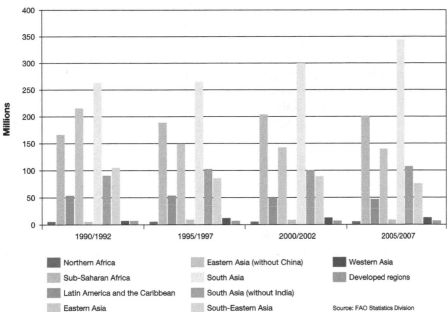

Fig. 6.2 Undernourishment in various regions of the world for various periods. *Source*: FAO (2010)

As illustrated in Fig. 6.2 there are still too many people that suffer from under-nourishment. In 2010, the regional distribution of people suffering from hunger was 578 million in the Asia Pacific Region, 239 million in sub-Saharan Africa, 53 million in Latin America, 37 million in North Africa, and 19 million in developed countries. These numbers may worsen as climate change, economic and political instability and disparities, persistently high food prices, and rapid population growth. Coming all at the same time, these factors would amount to a perfect storm for our societies and for our collective food security.

6.2.2   A New Food Equation

In terms of human well-being, the present food situation is characterized by a "double burden" of high numbers of people suffering from both undernourishment and over-nourishment. Given the precarious global water situation, this imbalance is serious and unacceptable. Sociopolitical inabilities to live up to pledges of "access to adequate food as a human right" result in considerable health and socioeconomic implications. In addition to the ones who are deprived of access to adequate food, the costs affect society at large such as in terms of health impairment and low ability to work. Like undernourishment, overnourishment has detrimental implications on human health, child cognitive capacities, and labor productivity.

There is one more significant component in the food equation, as discussed in Chap. 2, a large part of the food produced is lost before it reaches market outlets and another fraction is wasted at the end of the supply chain. The FAO estimates that 30–50 % of production is lost and/or wasted. These circumstances mean a pressure on water and other resources over and above what would be required to produce enough for a healthy diet. These circumstances also illustrate that a policy for food security must consider various dimensions and aspects from its production on the farm to the time it is beneficially consumed.

Given the increasing competition for water from other uses and the uncertainty and variability of precipitation (see Chaps. 2 and 3), it is necessary to produce "more crop per drop." Low levels of production and productivity are the result of several factors: insufficient capacity and inappropriate technologies in many parts of the world are especially problematic in combination with locally unfavorable production conditions. Wars and civil strife also result in low levels of supplies that do not meet demand.

The regional variation is significant. On wider geographical, economic, and geopolitical scales, an increase in the demand for animal products requires large quantities of feed. Demand for grains for feed purposes is in the order of 35–37 % of grain production (FAO 2009). This demand together with a conversion of another 5–10 % of the grain for production of biofuels push up the price on staple food items while also reducing supply of these commodities for direct human consumption. This might be positive for the producers but not so for poor consumers, who already spend a disproportionate share of their household budget on food. The unequal distribution of, and access to, food resources between people and countries is a harsh reminder of the injustice and, indeed, high social cost of poverty. It also suggests that the increasing pressure on water and other natural resources may not be correlated with increased human well-being.

If the world is having difficulty meeting global food security now, the challenges in the future loom large. This raises new questions about the food security on a planet with finite natural resources but with major imperfections in terms of sound stewardship of resources. The first decade of the twenty-first century saw several harbingers of a troubled future for global food security. The food price spike of 2008 with its consequent food riots and resulting political changes in several countries awoke the world's leaders to the reemergence of this threat to human well-being and social harmony (Nelson et al. 2010). Strong and new forces of change in the world food equation are transforming food demand, production, and markets.

The drivers on the demand side are discussed in Chap. 2. With income growth, globalization, and urbanization, demand for agricultural products will continue to grow and shift toward high-value commodities. The International Water Management Institute (IWMI) projections show that by 2050, under the business-as-usual (BAU) scenario, global grain demand will increase by more than one-third (34 %, 53 %, and 28 % for wheat, maize, and rice, respectively). These trends will be accompanied by very strong growth in meat consumption, especially in poultry and beef. Demand for milk is also likely to increase rapidly. These higher demands are explained mainly by the increase in the *per capita* income of the population in general.

These higher demands lead to increased prices for wheat (164 %), for maize (133 %), for rice (157 %), and other food commodities.

Despite this high demand and rising prices of agricultural products, production has been slow. The overall growth of agricultural products has been too low to cope with the fast growing demand. Between 2000 and 2006, grain supply increased by merely 8 %. There are short- and long-term causes behind this low level of production response. Of course, one of the primary long-term causes is the increasing competition for land and water and their rapid degradation. Production has also been impaired by decreased investments in agriculture. While the investment in agriculture grew annually at 1.1 % in the period 1975–1990, the rate was only 0.5 % during 1991–2007, according to data from OECD-FAO (2012). With declines in agricultural investment have come declines in productivity and sustainability.

6.3   Hunger for Land and Thirst for Water

The declaration of the World Summit on Food Security, held at the FAO in November 2009, stated that "to feed a world population expected to surpass nine billion in 2050, it is estimated that agricultural output will have to increase by 70 % between now and then." This 70 % estimate is now the most common figure used when predicting how much food production must increase in the next 3–4 decades. The calculations are based on projections of current trends in demand and population growth. They do not consider how the additional production will be accessible to those who most urgently need more food; e.g., small farmers, who are often net buyers of food, and urban consumers. If the cost of production and supply increases then, depending upon levels of subsidies, the poor will still face difficulties both to produce enough for them and to buy the food they need. In addition, the recommendation from the World Summit on Food Security does not consider the opportunities, which can be realized with a more efficient supply chain, i.e., by a reduction in losses and waste of produce. It is essential to discuss the realism of policies that aim at making the supply chain more efficient and to what extent the combination of strategies for increased production and reduced wastage are feasible.

Feeding the world's growing population and finding the land and water to grow the food continues to be a basic and sizeable challenge. It is an enormous task because the required increase in food production to meet future needs may have to be achieved with fewer land and water resources. Many countries lack the luxury of unused resources. Indeed, some regions face severe and increasing resource scarcity. South Asia and the Near East/North Africa regions have exhausted much of their rainfed land potential and depleted a significant share of their renewable water. More than 1.2 billion people today live in river basins where absolute water scarcity and the trend of increasing shortages are serious concerns. Expanding land under cultivation is possible in sub-Saharan Africa and Latin America but will require adequate farming practices, increased investments, and sustainable management of natural resources (Fig. 6.3) (Ait Kadi 2009a).

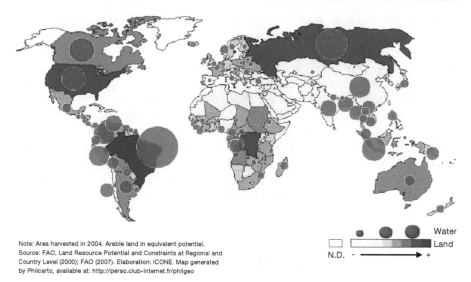

Note: Area harvested in 2004. Arable land in equivalent potential.
Source: FAO, Land Resource Potential and Constraints at Regional and
Country Level (2000); FAO (2007). Elaboration: ICONE. Map generated
by Philcarto, available at: http://perso.club-internet.fr/philgeo

Fig. 6.3 Available land and water for crop production (2007). *Source*: FAO (2007)

Using the WATERSIM model, IWMI estimates that the consumptive water demand at a global scale will increase from about 2,400 km^3 in 2010 to about 5,250 km^3 in 2050 for a BAU scenario. For an optimistic (OPT) socioeconomic scenario, the total consumptive water demand will increase to 7,230 km^3 by 2050, whereas for the pessimistic (PES) scenario the expected increase will not exceed 3,820 km^3 (Fig. 6.4).

Water availability for agriculture is a growing constraint in areas where a high proportion of renewable water resources are already being used or where transboundary water cannot be negotiated. Increasing water scarcity threatens irrigated production in some of the world's most important agricultural areas. In low to medium income countries with fast population growth, the demand for water is outstripping supply. Rising demand from both agriculture and other sectors is leading to environmental stress, socioeconomic tension, and competition over water. Where rainfall is inadequate and new water infrastructure development is not feasible, agricultural production may be constrained more by water scarcity than land availability.

Groundwater has provided an invaluable source of water for irrigation and other purposes, but has proved almost impossible to regulate. As a result, locally intensive groundwater withdrawals are exceeding rates of natural replenishment in key grain-producing locations. Excessive water pumping has made the groundwater levels in China, India, Iran, Mexico, the Middle East, North Africa, and the United States critically low. Because of the dependence of many key food production areas on

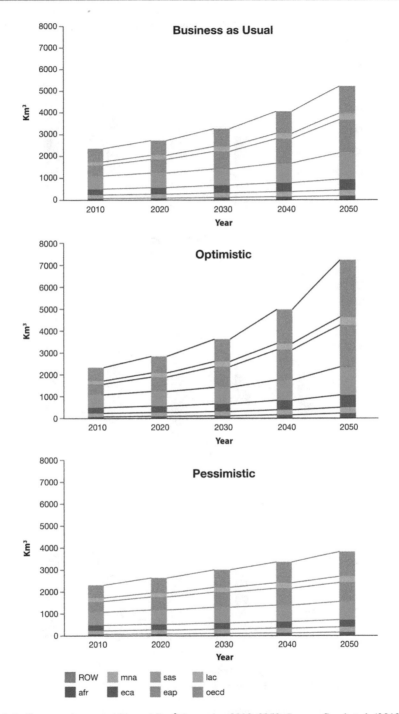

Fig. 6.4 Consumptive water demand (km³) by region 2010–2050. *Source*: Sood et al. (2013)

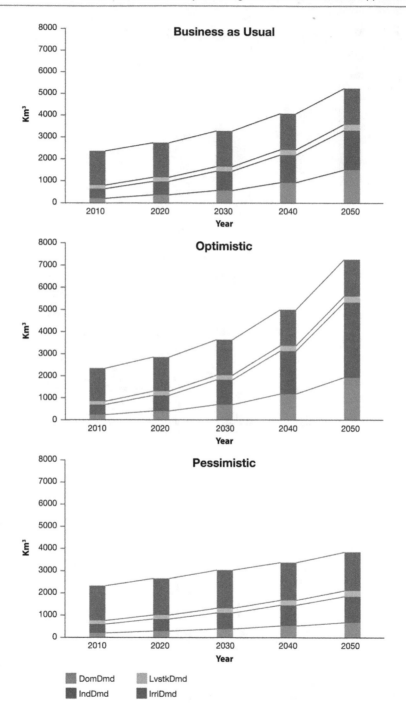

Fig. 6.5 Water demand by sectors under the three scenarios. *Source*: Sood et al. (2013)

groundwater, declining aquifer levels present a growing risk to local and global food production. In Yemen, e.g., groundwater withdrawals exceed recharge by 400 %, thus threatening the fundamental well-being of its citizens (Shah et al. 2000). With 25–27 million irrigation wells, groundwater irrigators of South Asia abstract over 300 km^3 of groundwater every year that provides supplemental irrigation to 70–75 million hectares of land. Private investments in groundwater wells have added more irrigated area to South Asia in the past 40 years than public investments in dams and canals added in the past 200 years. A booming groundwater irrigation economy is a unique aspect of South Asia's waterscape. It has become so central to South Asia's food security and agrarian livelihoods that its governments cannot afford to dismantle it. However, its environmental impacts are potentially so pernicious that they cannot afford to allow the groundwater boom to keep running amok as it has in recent decades.

In regions where natural resources are stretched, agricultural production has to compete with growing needs for water from other users. Growing cities, industries, and services have priority for water supply. Locally, this translates into reduction of the share of water available to agriculture. Municipal and industrial water demands are growing much faster than those of agriculture and can be expected to take water away from what otherwise would be used by agriculture. This, in turn, puts the ultimate source of naturally available freshwater—groundwater—under heavy pressure. Figure 6.5 represents the water demand by sectors under the three scenarios (BAU, OPT, and PES; see Box 6.1):

Box 6.1   The WATERSIM Model

IWMI developed an integrated water accounting and food trade balance model, WATERSIM, to look at future water and food situations.

For this analysis, the model was run until the year 2050, using base data from 2000 to account for the drivers of change. Three socioeconomic scenarios—i.e., high GDP growth with low population growth (optimistic, OPT), low GDP growth with high population growth (pessimistic, PES), and BAU—were considered along with two climate change scenarios (A2 and B1; Sood et al. 2013). The depletive water demand was separately calculated for domestic, industrial, livestock, and irrigation use. The depletive water demand is defined as the water lost due to evaporation and transpiration. The irrigation water demand was calculated as the difference between the potential evapotranspiration and the effective precipitation for each crop in the irrigated areas. The model optimizes the water that can be made available for consumptive water use given the available water resources (from a global hydrological model) while also maximizing the storage and meeting environmental flow constraints.

The optimization is done at monthly time-steps and at basin level (125 global river basins). If there is insufficient water available, the model prioritizes the

(continued)

Box 6.1 (continued)

water allocation in the order of domestic, industrial, livestock, and finally irrigation. The insufficient water availability for irrigation leads to reduction in crop yield and crop harvested area. This affects the food supply for the year and eventually trade. The global food trade optimization module of the model runs at annual time-steps for 115 global economic zones. The goal of food trade optimization is to bridge the gap between annual regional food supply and food demand (i.e., net trade of zero) by adjusting the world-clearing price for different crop commodities.

Demand for freshwater for cities and industries has doubled over the past 20 years and is predicted to increase by a factor of 2.2 from 900 km^3 in the year 2000 to 1,963 km^3 by 2050.

In other places, it is land, not water, which is the limiting factor for agricultural production. In large areas of Eastern and Southern Asia, including parts of India and China, demographics and demand for agricultural products are putting unprecedented pressure on limited resources. In parts of sub-Saharan Africa, in particular, Nigeria and Eastern Africa, land fragmentation has reached unsustainable levels, leaving farmers with cropping areas much below than necessary to ensure self-sufficiency. The expansion of urban areas and land required for infrastructure and other non-agricultural purposes is expected to continue, as we live in an increasingly urbanizing world. In China, for example, if in 2050 the entire population enjoyed a level of automotive equipment equivalent to the United States in the year 2000 it would require 13 million hectares of good farmland for roads corresponding to the needs of traffic, equivalent to about half the 29 million hectares currently producing 120 million tonnes of rice products to nourish the Chinese people!

The consequences of this "hunger for land and thirst of water" in the long-term are: (1) Asia and Near East/North Africa will be major importers; (2) sub-Saharan Africa could feed itself but with a low increase of *per capita* food ratio; (3) Latin America will be a major exporter (Brazil and Argentina) but with important ecological risks; (4) Canada and Russia could increase their export capacity; and (5) United States and the EU could increase also but in weak proportions.

6.4   The Century of Agriculture

The twenty-first century is the century of agriculture—because humanity will rely on agriculture to produce food, feed, fiber, and fuel. Indeed, it is not only more food that will be demanded. Millions of farmers are increasingly exposed to a growing demand for other commodities in addition to food (Ait Kadi 2009b).

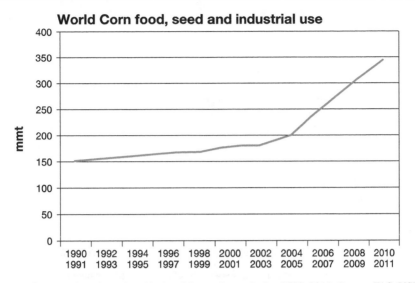

Fig. 6.6 Growth of food, seed, and industrial use of corn during 1990–2011. *Source*: FAS, USDA (2011) (production, supply, and demand online, Washington, DC, USDA)

The same water resources will also have to be used for other crops or commodities. The allocation of crops to nonfood uses, including animal feed, seed, bioenergy, and other industrial products, affects the amount of food available to the world. Globally, only 62 % of crop production (on a mass basis) is allocated to human food vs. 35 % to animal feed (which produces food indirectly as meat and dairy products) and 3 % for bioenergy, seed, and other industrial crops. A striking disparity exists between regions that primarily grow crops for direct human consumption and those that produce crops for other uses. North America and Europe devote only about 40 % of their cropland to direct food production, whereas Africa and Asia allocate typically over 80 % of their cropland to food crops. Figure 6.6 shows how food, seed, and industrial use of corn have expanded rapidly in the past 5–7 years.

The growth of crops for bioenergy has been highlighted as a potential competitor for land and water with food crops. According to the United Nations Environment Programme (UNEP), about 118–501 million hectares "would be required to provide 10 % of the global transport fuel demand with first generation biofuels in 2030. This would equal 8–36 % of current cropland, including permanent culture…" (UNEP 2009).

Worldwide there is rapid expansion of corn and vegetable use for fuel (Abbott et al. 2011). Many governments have set fixed mandates specifying the amount of biofuels to be produced, regardless of food and fuel prices. According to FAO (2008), biofuel production is projected to more than double from 2007 to 2019 and biofuel demand is expected to grow fourfold from 2008 to 2035 (IEA 2008). In addition, biofuel support is predicted to increase from USD20 billion in 2009 to USD45 billion by 2020 and to USD65 billion by 2035. Biofuel production has absorbed a rapidly increasing share of the US maize crop, for instance (Fig. 6.7)—today close to 35 % of the United States' maize production is used for biofuels!

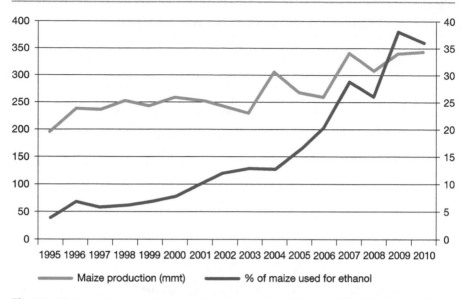

Fig. 6.7 Maize production and use for ethanol in the United States , 1995–2010. *Source*: Torero (2011)

The US National Academies of Sciences found that even if bioethanol production used all the maize and soybeans produced in the United States in 2005, it would only replace 12 % of the country's gasoline demand and 6 % of its diesel demand (FAO 2008).

The second large-demand growth category has been the use of oilseeds—vegetable oils for biodiesel production, oilseed meals for livestock production, and vegetable oils for human consumption. Figure 6.8 shows the increased use of oils for biodiesel and industrial usage. The percentage of total world use going to industrial and biodiesel usage has increased sharply since the mid-2000s. Rapeseed has been used extensively for biodiesel in Europe. For the world, nearly 33 % of total rapeseed is now used for industrial purposes compared to 17 % in 2004/2005. Industrial uses of world soybean oil expanded to 16 % by 2010–2011 from just 4 % in 2004–2005.

The impact of growth in world oilseed use is also led by the continued surge in soybean use in China for livestock feed, human vegetable oil consumption, and stock building in recent years. China largely abandoned its soybean self-sufficiency objectives, concentrating on self-sufficiency in feed grains, wheat, and rice (Fig. 6.9; Abbott et al. 2011).

The world's producers have responded to new demands by bringing more land into production and shifting from crops of low- to high-demand growth. Land area for 13 major crops increased by 27 million hectares since 2005–2006. This represents 3 % of harvested area. Twenty-four of the 27 million hectare expansion occurred in six countries or regions: China, sub-Saharan Africa, the former Soviet

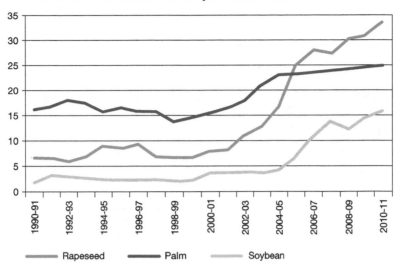

Fig. 6.8 World uses of oils for biodiesel and industrial purposes during 1990–2011. *Source*: FAS, USDA (2011)

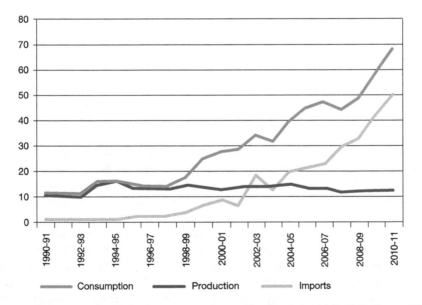

Fig. 6.9 Chinese soybean production, consumption, and imports during 1990–2011. *Source*: FAS, USDA (2011)

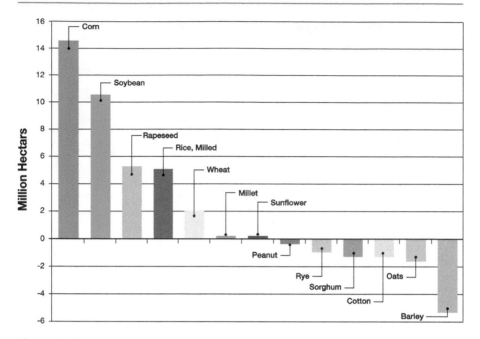

Fig. 6.10 Change in world harvested area for 13 major crops 2010–2011 vs. 2005–2006. *Source*: FAS, USDA (2011)

Union, Argentina, India, and Brazil (in order of importance). In addition, to meet growing demand, land shifted to high-demand crops of corn, soybean, and rapeseed, an increase of 30.5 million hectares worldwide. Rice increased five million hectares as nations encouraged expansion to reduce food insecurity (Fig. 6.10).

This process of equating the marginal return to land and water across crops will result in the future in high prices for many crops as water and land are the scarce resources. At the same time, the environmental benefits of biofuel production are being questioned.

Biofuel production with its effects on food price levels and volatility has contributed to the changing world equation. The International Food Policy Research Institute's (IFPRI) global scenario analysis until 2020 projects that biofuel expansion may result in price increases of 26 % for maize and 18 % for oilseeds compared to 2005 (von Braun 2008). As new linkages and trade-offs are created between the agricultural and energy sectors, agricultural commodity prices are becoming increasingly correlated to energy prices (von Braun 2008). The worrisome implication is that volatile energy prices will translate into larger food price fluctuations. Second-generation biofuel technologies, which may lessen the food–fuel competition, are still a long way away. Thus, how to meet ever rising demands for food, feed, fiber, and fuel while at the same time increasing farmer incomes, reducing poverty, and protecting the environment, all from an increasingly constrained land and water resources base, is a major challenge of the twenty-first century.

6.5   Climate Change Consequences

Global agriculture will have to cope with the burden of climate change whose likely impacts (higher temperatures, shifting seasons and more frequent and extreme weather events, flooding, and drought) have been documented in many reports. Most of them conclude that the global food production potential is expected to contract severely and yields of major crops like wheat and maize may fall globally. The declines will be particularly pronounced in lower-latitude regions. In Africa, Asia, and Latin America, for instance, yields could decline by 20–40 %. In addition, severe weather occurrences such as droughts and floods are likely to intensify and cause greater crop and livestock losses. Recent IFPRI analyses suggest that, calorie availability will not only be lower than in the "no climate change" scenario—it will actually decline relative to 2,000 levels throughout the developing world (Fig. 6.11). Climate change will also result in additional price increases for the most important crops: rice, wheat, maize, and soybean.

Climate change and global warming add to uncertainty and risk for food security that, for the chronically poor, will deteriorate in all its four dimensions as follows:

- Availability of food will probably be more uneven, with a likely decrease in some regions due to scarcity arising from declining water resources and worsening climatic conditions; changing food demands and a shift in the use of grains from food to feed and fuel production will alter supply of food, irrespective of production levels.
- Poor people's access to food will decline due to worsening terms of trade between wages and food costs and also due to higher prices on inputs in agriculture.

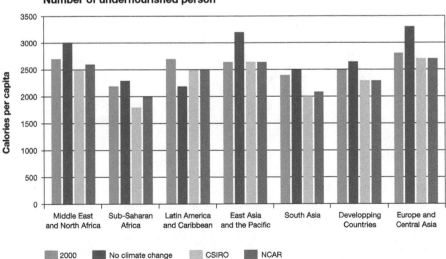

Fig. 6.11 Daily *per capita* calorie availability with and without climate change. *Source*: Nelson et al. IFPRI (2009)

- Stability of supply is threatened due to increasing prevalence of disasters, uncertainty regarding food prices, and national protectionism.
- Safe and healthy use of food will deteriorate as the poor switch to diets lacking essential micronutrients thus increasing child undernourishment. Increasing food insecurity might lead to more competition over water resources, migration, difficulties of supplying cities, and ultimately state failures and international conflicts.

Box 6.2 Consequences of Prolonged Droughts in Australia Especially in the Murray–Darling Basin 2002–2009 in Terms of Food Production and Farmers' Livelihoods

The southern Australian drought that lasted for most of the first decade of the twenty-first century significantly reduced the supply of some commodities. This drought was a contributor to increasing food prices that occurred during 2007–2008. The drought began in 2001. Following the drought wheat production, e.g., doubled (ABARE 2010). The effects of the drought were particularly profound in the Murray–Darling Basin that produces about 40 % of Australia's agricultural output. The National Water Commission funded the Commonwealth Scientific and Research Organisation (CSIRO) to undertake a modeling study of the basin aimed at determining the impact of climate change and development on water resources. The results emphasized the need for flexible management strategies (CSIRO 2008).

Water allocation trading helped irrigators manage variability in seasonal conditions. Typically, sellers of allocations received cash injections that helped them cope with drought and, in some cases, to manage debt. Purchasers maintained production or kept permanent plantings alive, thereby salvaging future production from long-lived assets. Water entitlement trading facilitates longer-term change. Entitlement trade volumes remained reasonably steady until 2002–2003. There were small increases in 2003–2004 and 2004–2005, followed by a sharp increase from 36 million m^3 in 2005–2006 to 388 million m^3 in 2007–2008. Sellers of entitlements turned to more opportunistic irrigation or ceased irrigation altogether. Purchasers have developed new irrigation methods or improved their reliability of supply.

Water trading in the southern Murray–Darling Basin added AUD220 million to Australia's GDP in 2008–2009; with net production benefits of AUD79 million in New South Wales, AUD16 million in South Australia, and AUD271 million in Victoria. Water trading has also contributed to securing critical urban water needs in Adelaide, Bendigo, and Ballarat. However, water trading may also have had some adverse consequences in that so-called *sleeper and dozer* entitlements (entitlements held by farmers who never or only infrequently used them) were sold and activated, thus increasing water extraction from the system (CSIRO 2008; ABARE 2010; NWC 2010; MDBA 2011).

6.6   Smarter Management of Water and Food Systems

Meeting the growing global demand for freshwater and food is a major global challenge. To meet this challenge we need dramatic improvements to every link in the human food chain. We need to look at the entire complex Web of issues characterizing the food system in a holistic manner. Despite the seriousness of the situation, the understanding of the water cycle-related linkages between different societal sectors is still weak. Goal conflicts remain unattended. Fundamental trade-offs need to be clarified in particular between the consumptive use of water involved in expanded food production and the water requirements of other sectors and ecosystems. The conventional, compartmentalized supply-oriented approach is not coping with the present water problems. Their solution requires an integrated approach to water, land use, and ecosystems addressing the role of water in the context of socioeconomic development and environmental sustainability. (Chapter 7 argues that energy production and use should also be in that mix of linkages to water.)

Fortunately, the successes and failures of the past 3 or 4 decades point us to other paths to avoid the peril. There is certainly no lack of authoritative global assessments that articulate pertinent water and food security challenges. Most of them respond, YES, it is possible to produce enough food to feed nine billion in 2050 but on certain conditions, including:
- Investments in agriculture, science and innovation, and social infrastructure to increase resources efficiency
- Constraining the demand for the most resource intensive types of food
- Minimizing waste in all areas of the food system
- Improving the political and economic governance of the food system and the functioning of markets at all levels
- Addressing climate change
- Reducing poverty

The Foresight study on the future of the global food and farming system undertaken by the UK Government Office for Science highlighted the importance of treating food production in a broader context, as one of major competitors for freshwater, land, and energy, and as integral to the world's overarching challenge of mitigating and adapting to climate change. Another Foresight study called Agrimonde, undertaken by INRA and CIRAD in France, has considered two scenarios to feed the world on the 2050 timeline: Agrimonde GO is a trend-based scenario that bets on economic growth to feed the world, in a context where environmental protection is not a priority; in contrast, the idea in Agrimonde 1 is to feed the world while preserving its ecosystems. Three of the clearest conclusions that emerge from Agrimonde are that: first, the food consumption patterns will have important impacts on the world food balance; second, major advances in sustainable food production and availability can be achieved through harnessing the progress of science and technology; and third, there is a need for a better functioning international food trade system (Foresight 2011; INRA and CIRAD 2010).

The Comprehensive Assessment of Water Management in Agriculture (CAWMA) concluded that the world has enough freshwater to produce food for all people in 2050 (CAWMA 2007). This can be achieved only if we act to improve water use in agriculture: 75 % of the additional food we need over the next decades could be met by bringing the production levels of the world's low-yield farmers up to 80 % of what high-yield farmers get from comparable lands. Better water management plays a key role in bridging that gap. The greatest potential for increased yields is in rainfed areas.

There is also a real scope to improve production in many existing irrigated lands. In South Asia, where more than half of the crop area is irrigated and productivity is low, with determined policy change and robust institutions almost all additional food demand could be met by improving water productivity in already irrigated crop areas. Similarly, in sub-Saharan Africa comprehensive water management policies and sound institutions would spur economic growth for the benefit of all.

Box 6.3 Meeting the Challenge of Growing Water Scarcity: The Example of Morocco

In Morocco, a consequence of increased industrialization and a rapidly growing population, accentuated by a progressive shift from rural to urban living, is a more intensive and comprehensive use of water resources.

The emphasis in Moroccan development planning for the last 3 decades has been on maximizing the capture of the country's surface water resources and providing for their optimal use in irrigated agriculture, potable water supplies, industrialization, and energy generation. Enormous capital resources have been invested in the essential infrastructure to control surface water flows. Infrastructure to capture and use about two-thirds of surface water potential is in place and a number of major infrastructure projects are in advanced stages of planning and/or construction to capture most of the remaining potential.

As the country nears the end of the infrastructure phase of its national development plan, emphasis is beginning to shift to the more sophisticated and difficult task of ensuring socially and technically efficient allocation of the existing water resources among competing consumer groups on a sustainable basis. This task is ever more complex given Morocco's relatively high population growth, the higher rate of immigration from rural to urban areas, and the great spatial and temporal variability in annual rainfall with droughts of frequent occurrence.

Despite remarkable achievements, Morocco faces a growing challenge in the water sector. One of the main issues is the decline in available water resources. The sustainable upper limit or "carrying capacity" of water resources use will be approached by the year 2020. Thus, a growing scarcity is anticipated as a result of rising demand due to expansion of irrigated areas

(continued)

Box 6.3 (continued)

and urban development and a slowing of the growth in available supplies, the depletion of aquifers, and the pollution of available resources. Per capita renewable water resources are expected to fall by 50 % in 2020, when all renewable resources are projected to be mobilized. At that time Morocco will move from being defined as a "water stressed" to being a "chronically water stressed" country. A number of river basins are already experiencing water shortages that will impose costly interbasin transfers. Some of the more intensively used aquifers are now considered to be under stress with serious drawdowns and saltwater intrusion in the coastal ones.

Increasing water resources development costs, along with severe financial constraints and competition for scarce public funds, have fostered a substantial change in attitudes to water conservation and serious questions have been directed toward water use efficiency and productivity in the irrigation sector, while recognizing its strategic role in the economic and social development of the country. Morocco's water economy is now characterized by sharply rising costs of supplying additional water and more direct and intense competition among different kinds of water users and uses. In this context, a better mix of supply and demand management is considered as the most rational response to water scarcity. Therefore, Morocco has adopted an integrated approach to water resources management through mutually reinforcing policy and institutional reforms. The major policy reforms adopted are the following:

- The adoption of a long-term strategy for an integrated water resources management. The National Water Plan will be the vehicle for strategy implementation and will serve as the framework for investment programs until the year 2020.
- The development of a new legal and institutional framework to promote decentralized management and increase stakeholder participation.
- Introducing economic incentives in water allocation decisions through rational tariffs and cost recovery.
- Taking capacity-enhancing measures to meet institutional challenges for the management of water resources.
- Establishing effective monitoring and control of water quality to reduce environmental degradation.

All these policy features are contained in the new water law promulgated in 1995. It provides a comprehensive framework for integrated water management. Some of the salient features of this law are:

- Water resources are public property.
- The law provides for the establishment of River Basin Agencies in individual or groups of river catchments. It clarifies the mandates, functions,

(continued)

Box 6.3 (continued)

and responsibilities of the institutions involved in water management. In particular, the status and the role of the High Water and Climate Council have been enhanced as the higher advisory body and as a forum on national water policies and programs. All the stakeholders from public and private sectors including water users associations sit in this council.

- The law provides for the elaboration of national and river basin master plans.
- It has established a mechanism for recovery of costs through charges for water abstraction and introduction of a water pollution tax based on the principle that "user pays" and "polluter pays."
- The law reinforces water quality protection by defining environmental mandates and enforcing sanctions and penalties.

Concerning the institutional setup, the major change is the establishment of River Basin Agencies empowered to manage individual or groups of river basins. The three principal responsibilities of these agencies consist of the development of water resources, the allocation of water as defined by the master plan, and the control of water quality. The agencies reinforce the network of existing institutions in charge of different water management functions.

Source: Ait Kadi 1997.

6.7   Toward a Viable Food Future

Today, humans are farming more of the planet than ever with higher resource intensity and staggering environmental impacts, while diverting an increasing fraction of crops to animal feed and biofuel. Meanwhile a billion people are chronically hungry. At the same time, unhealthy foods and diets are causing obesity, heart disease, and type 2 diabetes affecting two billion people. This situation is absurd and must not continue.

Is it possible to feed nine billion people? Long-term (2050) global food abundance is not totally guaranteed but there is no reason to be Malthusian and prophesy famines. We should, however, recognize that the constraints (reduction of available lands, water scarcity, increased risks of natural disasters, biodiversity loss, and social responsiveness) are so important that we have to consider a potential risk of temporary food scarcity (Ait Kadi 2009b).

Trends are not destiny. Changing contexts must be explored, opportunities identified in order to pave the road for better resources stewardship, more food security for all and per drop of water, and improved livelihoods for the producers as well as the consumers. In contrast with the deplorable trends in terms of the number of people suffering from food insecurity and the pessimistic outlooks of resource scarcity and increased uncertainty, it is vital to highlight that new opportunities are also emerging.

To meet the multiple objectives summarized under the preamble of this chapter, we need a holistic food system perspective. The potential threats are so great that they cannot be met by making changes piecemeal to parts of the food system. In this context, to meet the acute freshwater challenges facing us all over the coming decades, we need a collective new approach. We need to allow sustainable agricultural processes, systems, and technologies to be established. We need to ensure the sustainable use of water in other sectors. We need an approach that integrates the relevant parts of macroeconomic policies, agricultural policies, water supply and sanitation policies, trade policies, rural development policies, and environmental policies leading to institutional changes and infrastructure investments aimed at achieving a common goal.

For food security, we need to achieve efficient outcomes in all aspects of agricultural water management—from modernization of large-scale irrigation systems to enhancing water management in rainfed agriculture and better-linking livestock and fishery practices to water management. The water use sectors (including agriculture) that drive poverty reduction, economic growth, and development need to talk and work together. There needs to be high political ownership for that to happen. International agricultural trade should reflect economic and environmental comparative advantages and serve as a means to achieve global food security. We have to reduce losses of food in the chain from field to fork as well as the footprint and calorie intake of our diets.

Our way of life, our well-being, and our culture are intimately linked to how and where food is produced, what is produced, how we obtain it, how we prepare it, and how we eat it. The future of humanity depends on how food is—and will be—produced and provided.

Water and Energy

7

Water policy is much broader than providing enough water to drink, grow crops, preserve ecosystems, and reduce the risks of floods. It is also about ensuring sufficient water to produce the energy demanded by society. Humans and their economies and societies critically depend on reliable supplies of energy and water. Energy, as electricity and liquid and gaseous fuels, available when and where needed, requires water to produce. Water, of sufficient quality and pressure, available when and where needed, requires energy to produce. In short, energy is needed to provide much of the water we need and use, and water is needed to provide most of the energy we need and use. How can we ensure enough of both to meet all future water and energy demands? Limitations of either can constrain future economic and social development as well as adversely impact human and environmental health.

7.1   Understanding the Water–Energy Nexus

Everyone knows our planet is well endowed with abundant energy from the sun and water in and on our land and oceans. But to be useful where we live and work we must convert "raw" sources of energy and water to useful forms of energy and water. The important exception is the fraction of rainwater that infiltrates through the soil surface and becomes soil moisture, so-called *green water*. For virtually all other types of human activities, there is a close link between water and energy.

Energy must be in the form of liquid fuels or electricity, and water must be clean and under pressure. To produce electricity and liquid fuels, water is needed. To provide clean and pressurized water, energy is needed. Shortages of either can result in shortages of the other, and shortages of both can have adverse impacts on public health, the economy, and the environment.

Griffiths-Sattenspiel and Wilson (2009) estimate that 13 % of the total energy in the United States is consumed by collecting, treating, storing, and distributing water. Energy costs alone can account for about 75 % of the processing and distribution cost

Gulbenkian Think Tank on Water and the Future of Humanity, *Water and the Future of Humanity: Revisiting Water Security*, DOI 10.1007/978-3-319-01457-9_7,
© Calouste Gulbenkian Foundation 2014

of municipal water (Pate et al. 2007). In many cities, 30–50 % of the municipal energy budget is consumed in processing water. Yet few cities have sought ways to curb this expense through upgraded technology or conservation (AWWA 2011).

In California, the water-related energy demand consumes 30 % of all natural gas consumed in the state. This is enough to meet 60 % of all the household demand in California for natural gas. It consumes 20 % of all the electricity consumed in the state and is equivalent to the total electricity demand of the states of Oregon and Massachusetts. Managing California's water consumes more than 333 million m³ of diesel fuel per year (California Energy Commission 2005).

Conversely, the European Environment Agency calculates that the EU energy sector uses 44 % of the total water used in Europe. This is four times the amount used by industry and well above the 24 % needed for agriculture or the 21 % for public water supplies (Europe's World 2012, p. 99). Water's importance to the power sector was evident during the 2003 heat wave in Europe when French nuclear power plants had to shut down because the river water used for cooling was too hot (Europe's World 2012, p. 90). Table 7.1 lists this and other events where limited water has limited energy production in various regions of the world.

As the global population exceeds seven billion, with climate change exacerbating water stress in many countries, managing water–energy interdependences and their impact on climate will become an increasing concern at all levels of government and also in the private sector. The demands for both energy and water are projected to increase. This is driven largely by increased population and improving lifestyles that increase *per capita* consumption demands even without population growth. Figure 7.1 shows one estimate of the relative global trends in population, along with demands for energy and water.

The interdependence of water and energy, the "water–energy nexus," indicates the need to include applicable portions of the energy sector when analyzing water systems and applicable portions of water supply systems when planning and operating energy systems. This applies especially in those regions where demands exceed the supplies. An independent separate approach to managing these resources is likely to be less efficient and more wasteful than considering them together. This is especially so given that most of our useful water (except for soil moisture) cannot be had without the input of energy and most of our useful energy cannot be produced without water.

While making the argument that both energy and water production should be planned, designed, and managed as an integral system, we must remember that water and energy serve a multitude of purposes. Water for energy production and energy for water production are only a part of what those systems have been designed to do. All other purposes, such as water supplies for domestic and industrial uses, irrigated crop production, meeting the needs of livestock, recreation, and ecosystem quality, as applicable, need to be recognized when assessing how much water might be available for new energy development projects. The source and price of energy should be considered when planning and designing new water infrastructure projects. Recognizing this, we limit the scope of this chapter to the interactions between energy and water—acknowledging that in the "real" world, trade-offs among multiple purposes (and goals) abound.

Table 7.1 Events where limited available water lead to reductions in energy production

Year	Region/country	Climate event	Consequence
2001	Brazil	Drought	Combined with increased energy demand, the country experienced "virtual breakdown" of hydroelectricity and reduced GDP (Bates et al. 2008)
2003	Germany	Heat wave	Increased river water temperatures led German authorities to close a nuclear power plant and reduce output at two others (Cooley et al. 2011)
2003	France	Heat wave	Increased river water temperatures prompted the French government to shut down 4,000 MW of nuclear generation capacity (Cooley et al. 2011)
2006	Midwest, USA	Heat wave	Nuclear plants forced to reduce output at the time of peak demand. High river water temperatures, typically used for cooling, forced a Minnesota plant to reduce generation by 50 % (Averyt et al. 2011)
2006	Uganda	Drought	Hydropower reduced by one-third, with subsequent electricity shortages (Collier 2006)
2007	North Platte River, Nebraska and Wyoming, USA	Extended drought	After a 7-year drought, power generation from the North Platte River was reduced by about 50 %. A Laramie River coal-fired station, Wyoming, was at risk of insufficient cooling water and avoided impacts to power production by consuming water from local irrigation districts and the High Plains aquifer (Cooley et al. 2011; Averyt et al. 2011)
2010	Washington, USA	Low snowpack, followed by heavy rains	Given changes in precipitation regime, the peak stream flows were not aligned with power projections, straining hydropower generation and affecting electricity prices (Averyt et al. 2011)
2010	Lake Mead, Nevada, and Arizona, USA	Low water levels	Lake Mead water levels dropped to those not seen since the 1950s, prompting the US Bureau of Reclamation to reduce the Hoover Dam's generating capacity by 23 % (Walton 2010; Averyt et al. 2011)
2011	Texas, USA	Drought and heat wave	Farmers, cities, and power plants compete for the same limited water resource. After the driest 10 months on record (since 1895), at least one plant was forced to cut its output, and some plants had to pipe in water from new sources to maintain generation. If the drought continues throughout 2012, several thousand MW of electricity may go offline (O'Grady 2011; Averyt et al. 2011)
2012	Midwest, USA	Drought and heat wave	Braidwood twin unit nuclear power plant in Braceville, Illinois needed to get special permission to operate after temperature of cooling water rose to 102 °F (Wald 2012)

Source: Fencl et al. (2012)

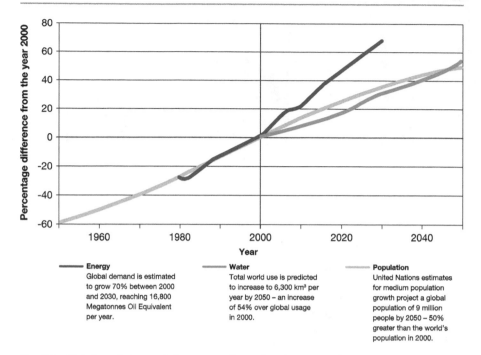

Fig. 7.1 Global trends in population, energy demand, and water use (*Source*: Edvard 2011)

Less water use means less energy use in extracting, treating, and transporting water. Less energy use means less water abstracted from rivers and lakes and groundwater aquifers. This leaves more water available for other uses including the environment and for maintaining aquatic ecosystem services. The net result of any saving may also be reduced emissions of greenhouse gases to the atmosphere. Additionally, there may be less discharge of polluted water into natural water bodies or that undergoes further treatment and possible reuse—that in turn requires more energy, and hence more water, as indicated in Table 7.2.

This interdependency between useful water and useful energy will be explored in more detail in the following sections.

7.2   Water for Energy

Water is used to extract fuels and produce from them nearly all forms of energy, including electric power and liquid fuels such as gasoline, kerosene, and diesel fuel.

A conceptual model is a convenient way to explain the interdependency of water and energy. First consider just the energy production and use components of this conceptual model, shown in Fig. 7.2.

To create electricity or liquid fuel a fuel source is needed. This is called a raw (or primary) energy source. After extraction and conversion processes to create electricity or liquid fuels, and then transport (or transmission), the (secondary) energy is

Table 7.2 Energy trends produce water use trends

Energy trend	Resulting trend in energy's water use
Shift from foreign oil to biofuels	Increases energy's water consumption if domestic agricultural irrigation water (and other inputs) is needed for fuel production
Shift to shale gas	Natural gas development using hydraulic fracturing may raise water quantity concerns if well development is geographically concentrated in areas with water constraints. However, natural gas from fracturing consumes less freshwater than domestic ethanol or onshore oil
Growth in domestic electricity demand	More water used for electricity generation. How much more depends on how the electricity is produced (e.g., smaller quantities needed if electricity is met with wind and photovoltaic solar, larger quantities if met with fossil fuels or certain renewable sources)
Shift to renewable electricity	Concentrating solar power technologies can use more water to produce electricity than coal or natural gas. These solar facilities are likely to be concentrated in water-constrained areas. Technologies are available to reduce this water use. Other renewable technologies, such as photovoltaic solar and wind, use little water
Use of carbon mitigation measures	Carbon capture and sequestration may double water consumption for fossil fuel electric generation

Source: Carter (2010)

Fig. 7.2 Conceptual model of the processes involved in converting raw energy sources to electricity or liquid fuels

in a form people can use. Raw energy sources include coal, oil, natural gas, nuclear material, geothermal waters, stored water in reservoirs for hydropower, the sun, wind, tidal and wave action and currents, and vegetation—including wood.

To convert raw energy sources to liquid fuels or electricity (secondary energy) two inputs are needed: water and technology. Each step of extraction, conversion, transmission or transport, and energy use sequence will cost money as well as consume water. For example, electrical energy lost in transmission will require more energy to make up for the loss, and this in turn requires water. The amount of water needed depends partly on the technology used to extract and convert the raw fuel as well as the means of distributing the resultant liquid fuel or electricity to where it is used.

Water is needed to drive the turbines at hydropower plants and water as steam is needed to drive the turbines of thermal power plants. Water is needed to cool both thermal plants and solar heat concentration facilities, for the production of biofuels, and to abstract oil and gas from belowground formations of shale. In short, water is needed to produce the energy needed by industry and the electricity that keeps our home and workplace air conditioners and furnaces operating, along with our stoves, refrigerators, washers, driers, radios, computers, cell phones, and television sets.

Raw fuel sources can be grouped based on whether they are renewable or nonrenewable. Among the nonrenewable fuels, crude oil or petroleum can be converted to kerosene, gasoline, and heavy fuel oil. Petroleum production requires drilling wells into oil-bearing formations and pumping the crude oil out. Water is needed for drilling and treating the crude oil. As the amount of crude oil pumped out decreases, extraction is improved by so-called secondary recovery processes that use water as steam to improve the viscosity of the crude oil and enhance pumping. Thermal steam injectors require water as well.

Another nonrenewable fuel source is coal. Open pit coal mining requires less water per volume of coal extracted than do underground mining operations. Extracted coal typically needs washing to remove nonfuel contaminants. The water not consumed in this washing process is clearly degraded in quality.

Natural gas, a third nonrenewable fuel source, is recovered by drilling wells into the underground layers containing the gas. Pressurized water or steam is pumped underground to force the oil and gas out of rock formations and to the wells. The water recovered in this process is often contaminated and presents a challenge to contain and control. Many of the promising, large remaining reserves of fossil fuels are water intensive to produce, including oil sands and unconventional natural gas obtained by hydraulic fractioning (fracking) in shale formations.

A fourth nonrenewable primary energy carrier is uranium. Uranium is present in the Earth's crust in the form of ores containing uranium oxide. It is recovered from open pit and underground mines and like coal, requires water for extraction. Additional water is required for milling, refining, and enriching the uranium.

Most of the energy produced in the world today comes from nonrenewable energy sources. A major portion of the world's electrical energy comes from converting raw energy sources to electricity in thermal power plants. These plants require water not only to produce steam but also for cooling. Cooling constitutes the vast majority of the water required by power plants. There is a range of cooling systems, but two types account for most of the power plant cooling. The first system (open-loop wet cooling) withdraws considerable quantities of water but consumes relatively little of what it withdraws. The second system (closed-loop wet cooling) withdraws less water but consumes a larger proportion of what it withdraws. Thus there is a trade-off between water withdrawal and consumption. Dry recirculating uses air as a coolant, so little, if any, water is needed (Feeley et al. 2008).

In the quest for increased sustainability and energy independence, the development and use of renewable energy sources are of increasing interest. Current attention is focused on biomass-based ethanol production, hydropower, wind and solar energy sources. Water needed to produce biofuels is mainly that required to grow the raw material, such as corn or soy. The amount of water consumed depends, in part, on the site as well as the fuel source.

Hydropower uses the potential energy of water to drive turbines generating electricity. Dams in rivers serve to store the needed water as well as to provide the water pressure (head) needed to produce hydroelectricity. The water losses associated with hydropower are mainly due to evaporation and seepage of water stored in the reservoirs. However, if the reservoirs serve multiple purposes in addition to hydropower production, the evaporation and seepage losses attributable to hydropower are not easily determined.

Energy directly from the sun, another source considered renewable, can be obtained from the sun's radiation. Solar energy can be used in three ways:

1. To heat water through solar collectors
2. To produce electricity from photovoltaic cells
3. To produce electricity through solar thermal power plants

These plants convert radiant energy into hot air or steam used to generate electricity and water is consumed in the process. Solar thermal farms use five times more water than nuclear power plants. In contrast, photovoltaic solar cells, which convert energy from the sun directly into electricity, use minimal amounts of water (USDOE 2006).

The conversion of kinetic wind energy to electricity does not require water, and the water requirement for turbine, tower, and transmission line construction is negligible. Both wind and solar energy systems benefit from some way to make up for the reduced energy production when the sun isn't shining or the wind isn't blowing. Hydropower reservoirs can serve that purpose as can gas turbine power plants—but at a cost.

Some 20 % of the world's electricity comes from hydropower plants whose turbines are driven by water flow under pressure, instead of steam, to produce electricity. The amount of hydroelectricity generated is far below its potential contribution, especially in Africa. Many regions where water is abundant (including parts of Brazil, Cambodia, China, India, Iran, and Laos) are currently expanding their hydropower production, driven by economic, energy security, as well as climate change concerns.

Increasingly some energy comes from solar and wind sources, and a small amount from the heat of geothermal waters and from tidal action and ocean currents. With the exception of solar heat concentrating systems that require water for cleaning and cooling, little, if any, water is consumed in the production of electricity directly from the wind and sun. So far, the electricity produced by these renewable technologies is a small fraction of the total energy demand (WEF 2009; Schumpeter 2011).

Electrical energy is expressed in power units over a period of time, e.g., kilowatt hours or megawatt hours, abbreviated kWh and MWh, respectively. Tables 7.3 and 7.4 summarize the approximate water requirements to produce a MWh of electrical energy or a liter of liquid energy (gasoline).

Liquid fuels provide the heat needed to run combustion engines. The British thermal unit, BTU, is a measure of the heat content (energy) provided by a liquid fuel. A kWh equals about 3,412 BTU. A million BTU is equivalent to about 30 L or 8 US gallons of gasoline (more precisely, a US gallon = 3.8 L).

Tables 7.3 and 7.4 and the paragraphs below provide comparisons of the water needed to produce a unit of energy depending on the source of energy and technology employed. The comparisons of water consumption per unit of energy, and energy consumption per unit of water, depend on the source of energy or water, and also on

Table 7.3 Water consumption requirements for raw fuel extraction and processing

Liquid fuels	
Method/source	Liters of water per liter of gasoline
Traditional oil extraction	0.1–0.3
Enhanced oil recovery	1.7–312.5
Oil sands	2.5–62.5
Biofuels: corn	312–3,625
Biofuels: soy	1,750–9,375
Coal	0.2–2.5
Shale gas	1.2–1.9
Electrical energy	
Source	Water quantity per MWh of electricity
Coal	5–70 gallons or 150–265 L
Uranium (nuclear)	45–150 gallons or 170–570 L
Shale gas	3–5 gallons or 90–150 L

Sources: USDOE (2008) and Carter (2010)

Table 7.4 Water consumption requirements for energy production from processed fuel

Liquid fuels	
Method/source	Liters of water per liter of gasoline
Oil refining	0.9–2.3
Bio-refining	
Corn ethanol	4–7
Biodiesel	3
Cellulosic ethanol	2–6
Coal	5–8
Shale gas	1.2-1.875
Natural gas processing	0.25
Electrical energy	
Method/source	Water quantity per MWh of electricity
Thermoelectric generation closed-loop cooling	190–720 gallons or 720–2,725 L
Nuclear	720 gallons or 2,725 L
Subcritical pulverized coal	520 gallons or 1,970 L
Supercritical pulverized coal	450 gallons or 1,700 L
Integrated gasification	310 gallons or 1,170 L
Natural gas combined cycle	190 gallons or 720 L
Thermoelectric generation open-loop cooling	100–300 gallons or 380–1,135 L
Hydroelectric	0
Geothermal	1,400 gallons or 5,300 L
Solar—concentration solar	750–920 gallons or 2,840–3,478 L
Solar—photovoltaic	~0
Wind	~0

Sources: USDOE (2008) and Carter (2010)

the place where extraction and conversion processes take place. Each single value is representative of a range of values, again depending on the place, and ignores uncertainties and errors in the data source. Hence these values only indicate the relative differences in the required amounts of these resources, identifying which options consume more or less resources, and about how much. Their exact values at a particular geographical location may well differ from the numerical values shown.

The volume of water consumed to produce a unit of useful energy depends on the fuel and technology used to produce it. For nonrenewable fuels, the demand increases from uranium to natural gas to coal, and finally to crude oil. The water demand for renewable sources varies from wind that requires only negligible amounts of water (if water stored in reservoirs is used to balance the energy load when there is no wind) to solar thermal sources, and hydropower, depending on the rate of evaporation and that depends on the climate where the hydropower reservoir is located (WEF 2009).

For biomass fuels, the water demand depends on the type of biomass (crop or plant material) used, the agricultural production system, and the climate that can affect any irrigation requirements. The water demand of average biomass grown in the Netherlands is less than half than that used in the United States and Brazil, and less than 20 % of that used in Zimbabwe. Based on the average *per capita* energy use in western societies, each of us uses about 35 m^3 of water annually to meet our energy needs. If the same amount of energy is generated through growing biomass, the water demand could be 70–400 times that of the other primary energy carriers (excluding hydropower). The trend toward larger energy use in combination with increasing contribution of energy from biomass will bring with it a need for more water. This causes competition with other water demands, such as water for food crops (Gerbens-Leenes et al. 2008).

The data summarized in Tables 7.3 and 7.4 show that, on average, coal- and oil-fired power plants consume twice the water required by gas-fired plants to produce the same amount of electricity. Nuclear power plants consume about three times more water than gas-fired plants and 1.5 times more than coal- or oil-fired plants. While costing more, integrated gasification combined-cycle coal-fired power plants reduce carbon emissions and consume similar amounts of water to gas-fired plants. However, they require about half as much water as do conventional coal-fired plants. Carbon capture from coal-fired power plants can consume 30–100 % more water due to reduced efficiencies.

For the same amount of energy, solar thermal technology today results in approximately five times more water consumption than does a gas-fired power plant, twice that of coal-fired power, and about 1.5 times that of a nuclear plant. However, technology is changing, including the use of dry cooling.

The World Energy Council (2010a, 2010b) estimates that in the next 40 years the water consumed to generate needed electricity will more than double. Electricity generation *per capita* is also expected to double by 2050, with the highest increases in Latin America, Africa, and Asia. However, the amount of water consumed to generate electricity *per capita* (due to technology improvements) is expected to stay the same or increase only slightly in Africa, Europe ,and North America but to almost double in Asia and Latin America. The modeling that is the basis for Chap. 6 predicts significant increases in the use of water by industry in the next 40 years, which includes power generation.

Box 7.1 Water Needed for Energy Development

In the aging oil wells of Saudi Arabia more water is pumped in to increase reservoir pressure than the amount of oil that is actually being pumped out. According to the U.S. Department of Energy, 2 to 2.5 L of water are used to produce each liter of gasoline from conventional crude and more than 6 L of water are required to produce 1 L of gasoline from oil shale. Alternative fuels are also water intensive. The voice of the U.S. ethanol industry, the Renewable Fuels Association, estimates that 3.45 L of water are used per liter of corn ethanol produced. Electric generation is no less water intensive. Ninety percent of all power plants in the US are thermoelectric, requiring billions of liters to cool the steam used to drive their turbines. In recent years, plans for new power plants had to be scrapped because water-use permits could not be obtained. In most countries in Latin America including Brazil, Paraguay, Peru and Argentina, hydroelectric power is the main source of electricity. Want to build a concentrated solar thermal power plant or a nuclear power plant which produce clean energy? Better make sure there is ample supply of water nearby. Solar thermal power plants require large amounts of water to create the steam that spins the turbines and for their cooling towers. Sunny places where solar power would otherwise be an ideal source of electricity often suffer from water shortages that make this form of energy a non-starter.

Source: Luft (2010).

7.3 Energy for Water

A conceptual model of the water abstraction, treatment, storage, and distribution processes is shown in Fig. 7.3. It is similar to the conceptual model for energy shown in Fig. 7.2.

Some water sources, such as rainfall, can be used directly for crop production. The amount of treatment, if any, may depend on intended water use.

Figure 7.3 identifies the stages needed to convert raw water sources to useful water, i.e., water available when and where, and in the quality and pressure needed. Raw water sources can be rainwater runoff collected and stored in cisterns, streams, rivers, groundwater aquifers, lakes and reservoirs, and the seas and oceans. Typically raw water must be treated and then stored and distributed before it becomes useful water. The amount of energy required to bring raw water to a treatment plant depends on the distance and elevation change. The energy required to pump groundwater depends on its depth below the surface. About 80 times more energy is required to lift each unit volume of water when depth changes from 35 to 120 m. Transporting a cubic meter of water 350 km horizontally requires about the same amount of energy as producing a cubic meter of desalinated water from seawater (Hoff 2011).

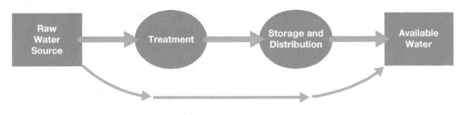

Fig. 7.3 Conceptual model of the conversion of raw water to water ready for use when and where needed

Table 7.5 Energy requirements for providing useful water

Process	kWh energy requirement per million gallons (or per 1,000 m³) of water
Groundwater pumping	140–540 (37–140)
Water treatment	
High-quality water	100 (26)
Brackish	1,200–5,200 (317–1,374)
Seawater	13,500–17,000 (3,570–4,490)
Wastewater	2,500 (660)
Distribution system pumping	Varies considerably
Storage prior to use	Varies considerably

The energy requirements of water supplies can vary regionally. In some areas, heating water for domestic use can use more energy than supply and treatment. In California, where water is pumped over long distances, transport alone can be the most energy-intensive process. The extent of irrigation is usually the most important difference between regions, since watering crops can require large amounts of energy depending on water sources, climate, and crop varieties (USDOE 2006).

The amount of energy required to treat raw water also depends on the incoming water quality and the required effluent quality, usually determined by water quality standards set by public environmental or health agencies. Finally, elevation change and distance that treated water travels to be where it can be used also requires energy. In short, energy is needed to operate the pumps and associated infrastructure that extracts, treats, delivers, and heats the water used in homes, businesses, factories, and farms. Table 7.5 lists estimates of the electrical energy required to convert raw to useful water. These costs can and do vary considerably depending on local conditions. Considerations are the quality of the source waters, groundwater depths, distance and elevation changes for transport, and local treatment plant characteristics and efficiencies.

Water transport, pumping, and use are energy intensive. Water is heavy. Energy is needed to lift, move, process and treat water at every phase of its extraction, distribution, storage, and use. In some areas water is pumped over long distances and elevation changes. Water heating in the United States can account for 14–25 % of the energy consumed in homes (USDOE 2011).

Pumping water out of groundwater aquifers requires more energy than obtaining water from surface supplies, again depending on the aquifer depth and surface topography. As a result of subsidizing the cost of electricity in India, there has been a significant increase in tube-well development at the expense of gravity irrigation systems (Shah 2009; Chartres and Varma 2010).

Energy is also required to recharge aquifers and recycle water for reuse. Pumping and injecting water into the aquifers consumes more energy and costs more money. If reverse osmosis is used for water recycling then the cost is high in energy. However, there are less expensive, less energy-intensive alternatives. The use of spreading basins, which allow water to percolate of its own accord down into aquifers, is less energy demanding, as is the use of wetlands rather than mechanical devices to filter water. Again, the options available, and their costs, depend on the particular site conditions.

Box 7.2   Energy-Efficient Centralized Wastewater Treatment in California

Flush the toilet and you may send around 9 L of water down the drain. Having a shower adds another 30 L or so, while a steaming bath contributes around 80 L. In total, the average person in the UK gets through around 150 L of water every day. But where does all this water go and what is the environmental impact of flushing it all away? For most of us our wastewater joins the labyrinth of sewer pipes under the city streets, and wends its way to the nearest sewage plant. For some people, however, particularly in more rural areas, their wastewater may not travel far at all, perhaps just down to the septic tank at the bottom of the garden.

These kind of small-scale decentralized systems are cheaper to install, cost less to maintain, use less water, and have greater flexibility in planning for future growth. For this reason they have become popular in recent years and are often seen as the "eco-friendly" option. However, a new study by Shehabi et al. (2012) indicates that centralized systems may have the smaller environmental footprint.

Arman Shehabi from Lawrence Berkeley National Laboratory, United States, and his colleagues directly compared energy use, greenhouse gas emissions, and air pollution at two water-processing plants in California—one centralized and one decentralized.

The centralized system serves around half a million people, over a 200 km² area. The decentralized system, meanwhile, treats the water for a community of 47 houses in Stonehurst, Martinez, in northern California. In this case each house has a septic tank, connected via a sewerage pipe to a small, local treatment plant.

Using a model known as the Wastewater–Energy Sustainability Tool (WWEST), Shehabi and his team were able to compare these systems using a

(continued)

Box 7.2 (continued)

full lifecycle assessment. They found that the economy of scale of the centralized system made it far more energy-efficient than the decentralized system. In their case, the centralized system used five times less energy than the decentralized system per volume of wastewater processed.

"The centralized system requires an enormous amount of infrastructure and operational energy in absolute terms, but the energy impact from this resource demand is reduced when normalized against the enormous volume of wastewater being treated," Shehabi told environmentalresearchweb (http://environmentalresearchweb.org).

In terms of energy efficiency, conventional centralized wastewater treatment systems appear to be the best choice. However, there are some advantages to decentralized systems. "Decentralized systems have the advantage of being able to be tailored to the specific needs of a small community," explained Shehabi. "They can be added incrementally rather than requiring a major public works project, and because of their smaller size are more amenable to wastewater separation strategies. This separation capability allows for water reuse, which becomes increasingly important in areas with limited water supplies, not only because water itself is a scarce resource but also due to the embodied energy of delivered water."

In an area where water is very scarce and sourced from energy-intensive desalination, e.g., decentralized systems are likely to have a significantly lower energy impact. Although this study was specific to California, Shehabi and his colleagues believe that the findings are applicable to many parts of the world. "Planners should not automatically assume that any decentralized system will be the low energy-impact option," said Shehabi. Instead they suggest that planners carry out a lifecycle analysis before installing new water treatment works, to ensure that any hidden impacts are revealed (see Chap. 5 for more on this subject).

Source: Shehabi et al. (2012)

7.4   Analyzing Water–Energy Systems

Figure 7.4 defines the interdependent water–energy system, and combines Figs. 7.2 and 7.3 and also includes water reuse. This figure shows the linkages between the energy and water systems.

This model also shows additional inputs of the resources needed for residual (wastewater) treatment and reuse possibilities, as applicable. Dotted lines indicate possible energy and water requirements, depending on particular situation. In some cases they do not apply.

This conceptual model showing the components of the interdependent water–energy system and their interactions can be converted to mathematical expressions quantifying relationships between energy, water, technology, and costs. These mathematical expressions can be used in analyses to define the marginal benefits of energy savings and lowered carbon emissions per unit of water conserved. Such analyses could identify the marginal benefits of every kWh of energy saved due to water conserved or allocated to another use. Conservation of both energy and water has benefits related to both resources that are often not considered. Analyses of the water–energy system can identify just how much energy can be saved if water consumption is reduced, and vice versa. Integrated water resources planning must include water use in the energy sector. Thus it follows that integrated energy planning must include water use for energy production (Braga et al. 2009; Perrone et al. 2011).

Decisions affecting future energy supplies need to consider water availability and costs just as decisions affecting future water supplies need to consider energy availability and costs. Both will impact our use of these resources as we grow our economies and protect our environments.

Considering and analyzing the water–energy infrastructure as an integrated system may identify ways of improving efficiencies and reducing costs, consumption, and emissions of pollutants. These, in turn, may positively influence the effect of climate change on public health.

Developing a model that captures all the components of a water–energy system as shown in Fig. 7.4 would permit analyses of coupled water–energy systems. Such analysis may not be of much relevance to local water utilities that view energy inputs as just an expense they have to pay and have an incentive to reduce if possible. Utilities can reduce energy costs by reducing energy use.

Similarly energy sector planners and operators typically have no authority or control over how water is provided, but view it as either a cost or a constraint or both. Therefore an initial step toward a more comprehensive water–energy system's analysis might be to develop models that contain only those components of the system relevant to particular water or energy providers.

Modeling energy–water systems to explore the impacts of options for decreasing costs as well as resource use and emissions will have little value unless institutional arrangements exist that can implement such options. Local and global leaders in both the energy and water industries will need to work together to identify and implement energy and water efficient solutions reflecting agriculture, economic, environmental, climate change, and other concerns of the twenty-first century. The challenge, of course, is that this will typically involve multiple institutions at multiple locations with multiple authorities that typically have little if any history of working together.

In addition to institutional challenges, adopting cost-effective management options that reduce both energy and water requirements are often hindered by other factors. These include limited staffing levels at water utilities, competing priorities at drinking water and wastewater facilities, and lack of public awareness about the impacts such options may have on the energy–water lifecycle. Options that may

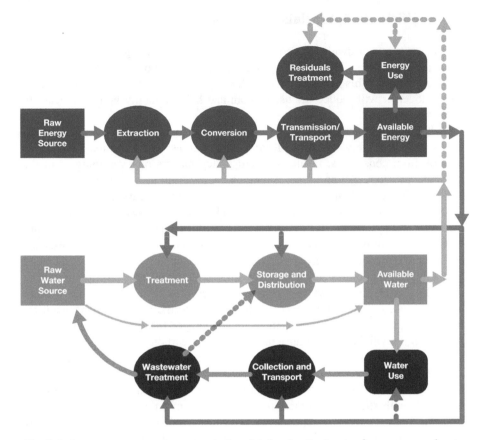

Fig. 7.4 A water–energy system conceptual model showing the inputs of raw energy and water sources needed to produce water and energy in useful forms

reduce costs include implementing monitoring and control systems, modifying pumping operations and pump efficiencies, modifying aeration operations, upgrading and right-sizing equipment, improving maintenance and leak detection technology, redesigning water systems, using renewable energy from solar, wind, and hydroelectric power plants as well as biogas from wastewater treatment plants.

The success of any technology and/or risk management tool depends on the local capacity to implement it. Many efficiency improvements require new operational management models and access to information that may not exist without further investments in monitoring and data management measures. These improvements could contribute to other objectives such as food security, reduced emissions affecting climate change, and improved public health, while simultaneously achieving water and energy use efficiency. Ongoing research and development may also lead to improved efficiencies in energy production.

7.5   Efficiency Potentials and Challenges

Many countries worldwide lack comprehensive water and energy development and management policies. Those that do have them typically show little if any cohesion between the two. This needs to change. A sharper focus on water and energy efficiency together will reduce the use of both and help humanity be more secure in many ways.

A holistic approach to water and energy takes short-, medium-, and long-term views of what society expects to have in their futures. This requires a change in outlook from consumers, businesses, and policy makers on everything from the price of water to recycling it and generating energy from it. It will not be easy, but getting it right now should save a lot of stress and effort in years to come.

Perhaps the most compelling reason for addressing the economic efficiency of the water–energy system is the broad reach both of these resources have in our economy and our lives. Water and energy supplies feed many competing uses, each essential to our social and economic well-being. Competitive uses of water extend throughout all economies and societies. Withdrawal options for power plants can be limited by needs to support in-stream water uses such as recreation and wildlife protection (GAO 2003). Future demands may add to the list of competitive uses. If carbon capture and storage is mandated or encouraged for fossil fuel power plants, these facilities will have to produce more energy for carbon sequestration. This will have corresponding increases in water withdrawals and consumption (Pate et al. 2007; NETL 2010; Atlantic Council 2011; Lyons 2012).

Developing and using renewable energy sources to meet increased demands and to comply with stiffer government requirements promoting reduced carbon emissions and more energy independence brings its challenges. Renewable energy sources, such as solar and wind energy, are not reliable all of the time. Utilities must assure a supply of electricity in two forms: energy and capacity. Capacity, the ability to produce dependable energy, is becoming more important as less dependable renewable energy forms an increasingly large part of the energy mix feeding into the grid. If solar and wind energy are eventually to generate a significant portion of the world's electricity, the industry must find effective ways of meeting energy demands when the sun isn't shining or the wind isn't blowing or when droughts adversely affect hydropower production. As the electrical grid depends more heavily on renewable sources, it will need other energy sources that can step in quickly to balance the system loads and supplies. Energy stored in batteries, in the form of water in hydropower reservoirs and in the form of heated water or salt—sometimes up to 15 h (Hoff 2011; WWF 2011)—are limited to satisfying relatively short-term shortages.

Unreliable electricity service leads to behaviors that waste both water and energy resources. In the agricultural sector, many farmers who pump groundwater use oversized motors to avoid burnout from poor electricity quality. Or they pump and operate supplementary fossil-fuel powered generators 24 h a day to compensate for irregular service. Urban residents with unreliable piped water systems run individual generators to pump water to rooftop storage tanks or to extract additional water from low-pressure municipal systems. Industrial consumers also invest considerable additional resources in backup power and water systems to ensure continued production levels. Moreover, a lack of emphasis on preventing water pollution increases the expenditure of energy either to treat water downstream or to access uncontaminated supplies from the ground or from distant surface sources.

In the consumption of water to generate electrical power, power plant inefficiencies or lack of mitigation measures often results in wasted water and energy, and greater degradation of water and other environmental resources. Significant water pollution problems can arise from inadequate environmental controls at all stages of the power generation cycle including fuel extraction (e.g., mining and oil and gas drilling), generation (e.g., acid deposition and dam construction), and plant waste disposal (e.g., fly ash). Understanding the key driving forces, feedback relationships, and effects associated with water and energy use and management can break the cycles of waste and environmental degradation. There are numerous opportunities to be both more efficient and sustainable in the use of resources through joint assessment, planning, and action.

The amount of energy used to produce useful water can be reduced as well. Installing more efficient pumps and motors in municipal water systems can cut energy use by 5–30 %, particularly when enhanced by supervisory control and data acquisition (SCADA) systems. Even more basic strategies such as locating power and water treatment plants near each other can curb energy and water demands. For instance, biogas from water treatment can be used to produce power, and if power and water treatment plants are located next to each other, or at least nearby, then the cost of transporting biogas fuel is reduced (USDOE 2006). These types of improvements, however, can be expensive and take a few decades to pay off (AWWA 2011).

Existing fuels can be replaced with less water-intensive ones. Natural gas combined-cycle plants use gas combustion in addition to steam generation to power turbines and so use half the water of traditional coal plants (NETL 2010). Alternative energy technologies can offer energy production with little to no water consumption. Solar photovoltaic and wind power both have this capability but require supplementation with other fuels because of poor energy storage capacities. In addition, some hydroelectric dams, aside from evaporation and seepage losses, have relatively low water consumption since water stays in-stream. However, they have a host of other impacts on river ecology and recreational uses (NETL 2010).

Transportation fuels also vary in their water demands. Although any of them can be fairly water intensive, biofuels should be considered with particular caution. Refining and processing of biofuels is comparable to that of traditional fuels. However, if irrigation is required to grow feedstock for biofuels, their consumption of water can be high. Even if they are rain-fed, they use water and land that could be used for food production.

Selecting appropriate technologies at various stages of water or energy production can also reduce costs. For example, although dry cooling systems at thermoelectric plants withdraw and consume essentially no water, they are expensive to construct and are hindered in hot, arid climates. Compared to closed-loop cooling, dry cooling can penalize plant efficiency by 2–25 % resulting in 2–16 % higher

Box 7.3   Water Constraints on Thermoelectric Cooling

More than 80 % of electricity in the United States is generated at thermoelectric facilities. Thermoelectric facilities can generally produce power as needed, according to consumer demand and fuel supply. This responsiveness to demand makes electricity from these facilities particularly attractive. Thermoelectric facilities can be fueled by a variety of fuels; coal, nuclear, and natural gas are the most common. Renewable sources such as concentrating solar power, geothermal, and renewable biomass also use a thermoelectric steam cycle. Thermoelectric power plants use fuel to produce heat to generate steam, which turns a turbine connected to a generator that produces electricity. Cooling is required to condense the steam back into boiler feed water, so the process can be repeated. With few exceptions, water is used to cool thermoelectric power plants in the United States. Thermoelectric cooling represents 44 % of the freshwater withdrawn nationally, but less than 6 % of water consumed.

The cooling options available for thermoelectric plants vary in their water withdrawal and consumption. Water withdrawal is the volume of water removed from a water source. Consumption is the volume lost, that is, no longer available for use. Excessive withdrawals can harm aquatic ecosystems, while excessive consumption depletes the water available for other uses. There are two common cooling methods: once-through and evaporative. Once-through cooling pulls large quantities of water off a water body, discharges the power plant's waste heat into the water (which typically raises its temperature 10–20 °F), then returns the majority of the withdrawn water. Once-through cooling, while largely nonconsumptive, requires that water is continuously available for power plant operations. This reduces any opportunity for this water to be put toward other water uses and can make cooling operations vulnerable to low streamflows. Evaporative cooling withdraws much smaller volumes of water for use in a cooling tower or reservoir, where the cooling water dissipates waste heat by evaporating. Evaporative cooling consumes water. Many power plants operating where water is relatively plentiful use once-through cooling. The majority of power plants in dry regions use evaporative cooling, although some coastal facilities use saline water for once-through cooling. In general, older thermoelectric plants use once-through cooling. Withdrawal in once-through cooling affects the ecology and quality of the water body (e.g., elevated temperature and chemicals of the discharged cooling water). This has resulted in newer power plants generally using evaporative cooling.

Source: Carter (2010)

costs (USDOE 2006). Closed-loop systems can be advantageous in areas lacking abundant water supplies to support high withdrawals. Unfortunately, these are also often the same places where the higher consumption of a closed-loop system can be more devastating to declining water levels and reduce the water available to competing uses. These trade-offs are meaningful to power plant developers. Thus water intensity and availability can be major factors that influence decisions on where to site a plant and which cooling system to use (GAO 2009).

Consumers changing their behavior can also drive cost reduction from conservation. But changing how we all behave is not as easy as spending money on improved infrastructure. Economic incentives that promote more efficient use of some resources can motivate behavior change. Stronger rules and economic incentives motivating the improved stewardship of common, transborder resources—such as water—may help prevent the overuse of available resources to the point of their destruction. Continued investment in technologies and infrastructure that increase the efficiency of resource extraction, distribution, and use is also a possible approach to improved efficiency.

Voinov and Cardwell (2009) note that management of demand is largely neglected in practice, even though demand in part determines the available supply. They maintain "curbing demand is cheaper, faster, and ultimately more beneficial to individuals than increasing supply." Conservation can also be promoted by using regulation to price water to reflect its true value when in short supply and to account for the cost of treating low-quality supplies (Atlantic Council 2011). Increased costs and behavioral changes on the part of both consumers and businesses can reduce demand. The International Energy Agency notes that subsidies for fossil fuel are five times higher than those for renewable energy. But as for many subsidies, phasing out this support is difficult because the fossil fuel economy is so entrenched (Sills 2011).

Dealing with the inconsistencies of water pricing will be no less challenging. It is difficult to tell voters that their right to water is limited. Water is vital for life as well as being a key factor in every sector of the economy. Access to it is seen as a human right. Pricing water at less than what it costs to deliver discourages people from using water efficiently, from innovating to reduce water use, and from thinking about the risks of disrupting supplies. The same applies to energy supplies.

The perception of the value of water and energy can further influence the management of both of these resources. When supplies are in decline, even small decreases in supply will result in very high price increases (Voinov and Cardwell 2009). For example, cities that subsidize the cost of municipal water supplies to keep water prices low conceal the true value of the water. Raising water prices, while politically unpopular, encourages conservation (Atlantic Council 2011; AWWA 2011). As water supply issues continue to emerge, consumers cannot expect water to remain cheap (Webber 2008). The high value of water is becoming apparent as more water managers find it cost-effective in the long run to implement expensive wastewater reclamation and desalination projects.

7.6   Water, Energy, and Climate Change

Water and energy are linked to climate change through the emission of greenhouse gases. The use of fossil fuels for electricity, heating, and transport is a major contributor to higher greenhouse gas emissions. The close relationship between energy production and greenhouse gas emissions has led to the understanding that water and energy should be addressed not only as a single interdependent system but also having links to climate. One part of this challenge is quantifying the relationships among energy, water, and climate. Other chapters in this book remind us we should also include agriculture and the economy in general in this integrated system.

Once again, emissions from water–energy systems can impact our climate. Changes in our climate will doubtless influence water and energy availability and demand. Long-term shifts in water and energy supply will be subject to already emerging changes in precipitation, seasonality, timing of snowmelts, and temperature patterns, among other factors. Effects will vary by region but will be of particular concern in arid and semiarid places with high-energy demands (USDOE 2006; Pate, et al. 2007). The combination of an intricate and far-reaching network of influence, population, and resource demand increases, and the consequences of climate change make water–energy interdependence a central but complex issue. This begs for the attention of the entire community of social and economic stakeholders, administrators, and policymakers.

Today it is clear that changes in the climate extremes are probably exacerbated by the emission of more greenhouse gases, a result of increases in population and economic output. Societies need more energy for development, but most current energy technologies need water. Water use and availability and energy production have an impact on climate and changes in climate affect the availability of water. The availability of water, in turn, has an effect on energy production.

Climate change is likely to alter electricity production and use, ultimately affecting water. More power will be needed, and more water withdrawn and consumed to offset reductions in the efficiency of power plants and of transmission and distribution of electricity in warmer temperatures. Reductions in hydropower generation and increases in electricity demand associated with warmer temperatures will increase demand for new power generation—leading to increases in water withdrawals and consumption. Climate change mitigation policies, such as increased use of renewable energy sources and implementation of carbon capture and storage, may alter the demands on water resources. These impacts are not typically integrated into current energy analyses and policy (Arent 2010).

Reducing energy use in water management is not currently a major strategy for addressing climate change. Pumping water from its sources, through treatment processes, and then within distribution systems to the taps of customers in their homes and workplaces is among the larger uses of energy. If such energy use could be reduced quickly and significantly on a global scale, it could have a beneficial impact on our changing climate.

If present climatic trends continue, about one billion people in snow-fed river basins will have less water. There will also be more extreme weather events—frequent and violent floods and harsher droughts. The IPCC (2007) warns that yields on rain-fed lands in Africa may be halved. This threatens community survival, risks state collapse, and increased migration. Many regions may also experience changes in the periods in which water is available. This will complicate energy production planning. What may all this mean for water and energy security in the future?

7.7   Water and Energy Security

A rapidly rising global population and growing prosperity are putting pressures on resources. Demand for water and energy is expected to rise by 30–50 % in the next 2 decades. Simultaneously, economic disparities favor short-term responses in production and consumption that undermine long-term sustainability. Shortages could cause social and political instability, geopolitical conflict, and irreparable environmental damage. Any strategy that focuses on one without considering the other risks serious unintended consequences (WEF 2011, p. 2).

Box 7.4   Droughts: Energy Disasters in Slow Motion

"What we are seeing now is the way we produce electricity can threaten our water supplies, and it's already compromising our water quality across the nation," said Wendy Wilson, director of River Network's energy and climate program and author of the new report "Burning Our Rivers." These results echo a report released in Fall 2011 by the Union of Concerned Scientists (UCS), which also found that conventional power plants are stressing the United States' lakes and rivers by removing too much water or discharging it at extremely warm temperatures, to the detriment of surrounding ecosystems (Pyper and ClimateWire 2011).

In the southeastern states, which have been battling a drought for more than a year, the impact of power plants is especially worrisome and could lead to brownouts and blackouts throughout the summer and beyond. "The conflicts between energy and water needs are ones we've seen before…and will only worsen as the frequency of drought increases and water temperatures rise driven in part by climate change," said Ulla Reeves, regional program director at the Southern Alliance for Clean Energy. Alabama along with other states in the southeastern portion of the United States is currently looking at all options for managing its fragile water–energy nexus and to meet the deadline given for a comprehensive water plan by 2013. According to the program director at the Alabama Rivers Alliance, the state has no official water management plan to ensure that users get the water they need while enough is left over to protect the river network itself.

"Efficiency in our water use and making sure that we're not wasting water or wasting electricity, which, in turn, uses more water, has got to be addressed in part of a comprehensive plan," said the program director.

The nation needs a plan, too, said Wilson, if it is going to weather the existing and expected water shortages. "It's not clear that the water and energy agencies are in a process where they are talking about these scenarios and building the resilience that we need to face these kinds of challenges," she said. "Is someone managing the farm here on integrating our water and energy needs? Or are we just praying for rain (Pyper and ClimateWire 2011)?"

Ample water is available on this planet to meet all human and environmental needs. However, there are places and times where the demands for acceptable quantities and qualities of water exceed the available supplies. Engineers know how to treat and transport water from one place to another anywhere in the world, but at a cost in infrastructure, the environment, and energy. Examples range in scale from the infrastructure needed in any urban area to provide treated water at every tap in every home, school, hospital, and industry to regional aqueducts that transport water from regions of plentiful water to water-scarce regions. Examples of the latter include transfers of water through the California aqueduct from northern to southern California, from beneath the Sahara Desert of central Libya to their populations along the Mediterranean coast via the "Great Man-made River." And also from the rivers in southern China—the Yangtze, Yellow River, Huaihe River, and Haihe River—to the drier north via the South-to-North Diversion Canal.

Energy is also used to collect and treat and possibly reuse water that has already been used, perhaps multiple times. But where and when there isn't sufficient water to meet all water demands at acceptable costs, there may be shortages in energy. This may be due, in part, to a shortage of water required to produce it. This situation exists today, especially but not only in regions of Africa, Asia, and Australia. As indicated in Table 7.1, water shortages have also affected the energy available in parts of the United States (Cooley et al. 2011). Furthermore, without significant changes to the ways we produce and consume, we will demand about 40–50 % more energy than we use today by 2035 (Hoff 2011). Can the supply of energy and water keep pace? Can these projected demands be reduced by various demand management measures?

Reliable supplies of water and energy are essential for economic growth and social stability. Economic and population growth increase both water and energy demands. Improving living conditions lead to more resource-intensive consumption patterns. Environmental pressures also drive resource insecurity—from climate change to extreme weather events that alter demands for both water and energy. Warmer climates can only lead to higher demands on both useful energy and water. Evaporation losses will increase. This will lead to higher water losses associated with hydropower and higher irrigation demands, which in turn will involves more pumping. Add to this the increased energy demand for air conditioning and other services expected as people's lifestyles improve despite, if not because, of improved energy efficiencies.

Unless offset by advances in technology, sustained increases in commodity prices and shortages of energy and water will limit economic growth. Should this happen, the poorest will likely be affected most, increasing economic disparity and the risks that this implies. Civilization will eventually be forced into switching to alternative fuels, and the end result will come down to money. When energy and/or water become more expensive, people will use less of it.

Energy production technologies and trends in energy consumption have substantially changed over the last century. Rapid population growth, increasingly energy-demanding lifestyles, and the exploitation of energy resources from fossil fuels to nuclear power have all helped to shape our energy environment. They have also prompted the need for a closer look into the economic, environmental, and security implications of energy

> **Box 7.5   Does Increasing Energy Efficiency Save Water?**
>
> In the 1860s in Britain, William Stanley Jevons observed how increases in energy efficiency resulted in higher energy consumption. Modern terms for this idea include "rebound," which refers to any increases in energy use that diminish the benefits of using those more efficient machines, and "backfire," meaning the increases go beyond 100 % of previous use. As appliances, cars and industry use less energy, people's behavior will tend toward finding ways to use even more energy. Owen (2010) quotes economists who are studying this anew finding that the Jevons Paradox rings true even though Jevons was talking about coal in the nineteenth century. Today we're talking about coal but also about natural gas, petroleum, and nuclear power. When implemented in isolation, technological solutions are more likely to result in rebound of various forms. However, by establishing people-centered approaches to energy and climate and by instituting smart policies, much of the rebound effect can be mitigated (Ehrhardt-Martinez and Laitner 2010). How society will reduce its ever-rising energy consumption in the face of finite supplies and increasing demands poses both challenges and opportunities. This suggests that it could be beneficial to increase our efforts on energy demand management along with energy supply management as we do with regard to water supplies and demands.
>
> *Source*: Woodside (2011).

alternatives and the increasing demand these alternatives are placing on water resources and the ways in which water quality and abundance are being considered or neglected.

It is natural to question the advantages of nuclear technology in the wake of recent events in Japan. Yet policy makers, environmentalists, and individuals should not hastily reject this carbon-free energy source. This is especially so when one considers the financial, safety, and ecological risks associated with alternatives like coal mining and deep-sea drilling. Issues currently confronting the nuclear industry in addition to safety or security (and lack of public trust) are costs, partly because of safety and security concerns, and radioactive waste disposal. We still have no way of economically denuclearizing radioactive materials. So we tend to think of long-term (really long-term) storage. Effectively addressing these issues could substantially reduce the political resistance toward implementing this technology that could lower greenhouse gas emissions—and to some extent, water consumption.

Contamination of water used for energy production is another major risk and concern. Contamination can result from faulty infrastructure designed to prevent direct contamination but also from the processes used to extract energy such as oil and gas. Older oil and gas fields that have had water injected into them under pressure often produce more water than oil or gas. This water is usually too contaminated to release into the environment without damaging it. Yet such polluted waters are often discharged into water bodies without treatment just to reduce the costs of oil and gas extraction. Acid drainage from coal mines and fertilizer runoff from crop production for biofuels may also contaminate freshwater resources.

7.8   Water and Energy Governance

The challenges associated with managing trade-offs involving the allocation of energy and water resources typically rest with governments. Many of the current trends in energy and water demands are in part driven by governmental policy. In many countries federal governments set the energy policy (often ignoring the water requirements) and at times define a vision for their nation's energy future (e.g., setting biofuel or solar energy production targets). Without integrated policy analyses involving all physical resource inputs, including water, the result-ing policies may be uninformed at best. At worst they may be inefficient or infeasible in the long run. Because affordable freshwater is a finite resource, commitments of water supplies for the energy sector reduce their availability for other sectors and for ecosystems. "Local or regional competition for water is often what makes energy's water demand significant; at the same time, it is the regional and local scale of water resources and how they are managed at those spatial scales that often complicates many national water-related actions. If energy security is a national security issue, is energy's water use by association a national security issue? Would this be a justification for federal spending on energy and water efficiency measures?" (Carter 2010).

Governments typically have separate administrative structures and independent policies for water, energy, and development, making coordination among them challenging at best. In addition, governments are increasingly under pressure to pursue multiple goals of enabling economic growth, reducing greenhouse gas emis-sions, and promoting energy and water use efficiency. The creation and use of com-missions that cut across government departments, stakeholders, and country representatives could lead to more effective management of both water and energy. One clear message is that energy policy must be developed in a way that carefully balances water availability and quality, national security, and financial costs and benefits. And this is all without losing sight of the short- and long-term environmen-tal effects of these decisions (Glassman et al. 2011).

An example of a regionally focused, integrated approach to water and energy use is defined in the Mekong River Commission's Strategic Environmental Assessment. This assessment has been able to consider the interdependencies between energy generation and water availability. Additionally, it has considered the impacts of alternative policies for energy and water resources development and management on ecosystems, social systems, and economic development over a 15-year perspec-tive (MRC 2009). Again, its implementation requiring cooperation and coordina-tion among the basin's countries is far from being assured.

Planning infrastructure development to meet increased energy demand is diffi-cult enough in an energy resource constrained world. It becomes more difficult when water resource constraints are considered as well. No one governmental or private organization has a monopoly on how best to develop more reliable and cost-efficient energy and water delivery systems. No one public or private organization is in a position to define for the society in which it functions what objectives should be achieved and the best way to achieve them. Whatever the objectives are as defined by any consensus among stakeholders, they will change over time, as will the best way to achieve them. Planning efforts must be comprehensive, and energy and water industries must be adaptive, considering the limits of both types of resources and financing. The growth of the region's population and modernization will surely lead to increased energy and water demands (Hajer 2011).

Economic disparity also often exacerbates risks as governments and consumers seek short-term, unsustainable solutions to economic hardship. An example is growing high-value, water-intensive export crops in water-deprived regions. Incentives now in place to use more biofuels in the United States are estimated to require up to the current amount of water now used for agriculture (NRC 2008). This is clearly unsustainable. Shale gas extraction promises access to new reserves of natural gas. But it poses a risk to water quality as well as possibly contributing more to greenhouse gas emissions than what it would reduce by replacing alternative fuels (Howarth et al. 2011).

Resource management can ultimately benefit from better communication and coordination among energy and water utilities and managers and the many governmental agencies involved with water management. Issues to address can include more flexibility in implementing environmental regulations, better coordination when managing interstate and international water rights agreements, and better and more frequent consultation on water rights (GAO 2003). Improved data collection and coordination can result from developing personal, working relationships between utility or agency managers (Goldstein et al. 2008). Proprietary issues and other barriers to trust can arise less frequently when stakeholders are personally familiar with each other.

7.9   Water and Energy: Future Perspectives

It is not hard to imagine a future in which population and economic growth will increase. This in itself will increase the need for more water and energy, but so will the increasing needs of agriculture, industry, the environment, and public supply.

The rising demand for water and energy from agriculture and from fast-growing and industrializing cities in Asia, Africa, and Latin America increases the challenges placed on water–energy systems. The global population in 2050 is expected to be two billion more than it is today. Greater wealth in many emerging markets is resulting in new and growing cities and increased, and more water intensive, consumption. The inhabitants of the new cities will likely consume more water than their forebears and to do this they will need more electricity. This of course could increase global carbon emissions. The impact on our ecosystems and our health is currently unpredictable.

If we are to be dependent on hydrocarbons for a large share of our energy for the next several decades, we need clean coal technologies and effective carbon capture and storage mechanisms. But current technological solutions for both are water intensive, and the water needed may not always be available where and when it is needed. Projected water demands for primary fuel extraction, carbon sequestration, and alternative transportation fuels derived from biofuels, shale gas, oil sands, coal, hydrogen, and the development of natural gas supplies from shale gas are significant. Likewise efforts to improve water security through construction of desalination and/or wastewater treatment infrastructure and/or large inter-basin transfer projects have significant energy implications.

Shifts toward cleaner energy—such as the increased use of biofuels and coal with carbon sequestration—can significantly increase freshwater demands. Existing hydrocarbon and alternative energy production is largely evolving toward greater water intensity. This can lead to shortages in energy due to water scarcities or high temperatures that limit cooling. Despite these concerns, water and energy policies are rarely integrated. "The disconnect between water and energy policy is driven in large part by the failure of water and energy practitioners to engage with and fully understand one another" (Cooley et al. 2011). Part of the reason for this is the geographic scales of concern to energy and water utility managers are usually quite different. Energy providers are rarely focused on regions as small as a city, or town, or county that water utility managers are responsible for. And water utility managers of local municipalities are not likely to feel they need to take into account the production of electricity or gasoline hundreds of kilometers away that they may eventually use.

While populations are expected to rise significantly in the coming decades, accessible freshwater supplies are not. Moreover, population movement and energy demand do not always track well with water availability. For example in the United States during the 1990s the largest regional population growth (25 %) occurred in one of the most water-deficient regions, the southwest. The situation in many developing countries is similar.

Sustainable energy and water development and production in the future will require an understanding of the interdependencies of both components of such systems. A balance among the needs of all users of energy and water and an appreciation of the impacts on climate, the environment, and the economy will also be necessary. This will likely require some changes in human behavior.

In a world of limited energy and water resources and where more people want and can afford more of them, shouldn't we examine our current management policies and uses? In particular, could we produce more energy with less water? The science is available, but too often the incentives for water use run backwards—to increase the demand for water and not to foster change. International trade policies do not encourage economical water use (van der Veer 2010). Couldn't we stop using drinking quality water for fighting fires or for flushing toilets and transporting wastewaters in urban areas? Addressing these and similar questions can help us in our attempts to achieve a more energy and water sustainable future.

Water Projections and Scenarios: Thinking About Our Future

8

The world faces serious water management challenges. Current planning and management approaches, technologies, and institutions have been insufficient to address the challenges of meeting future human needs for water, maintaining and even improving agricultural productivity of water use, achieving energy objectives, and satisfying growing industrial water requirements. These goals can and must be achieved while still protecting water quality and the biodiversity of vitally important natural ecosystems. We all need to work toward realizing this or other positive futures. We must debate what futures we might want, and then define and implement measures to move us in the right direction, toward those visions. Otherwise, business-as-usual approaches will lead us to a future we know is unsustainable and humanity will be much the worst off.

8.1 Thinking About Our Water Futures

Scenarios of the future are important tools for water planners and managers. If we cannot "see" the future, or possible futures, how can we adequately plan or prepare for them? Traditional water scenarios have included both qualitative (narratives or storylines) and quantitative (numerical data, tables, and graphs) assessments. These are typically based on current and projected water supplies and uses and on various assumptions and drivers that will influence what we might observe in the future.

Many water scenarios were produced during the period 1990–2000. There is also some literature available on global water scenarios extending farther back (for a review, see Gleick 2000a). This chapter describes the major characteristics of water development and uses scenarios including the timeline, scale, input, and output data, and major drivers and assumptions. Some details are provided on the role of water policies and strategies in influencing ultimate outcomes. Our assessment follows two steps: first, a literature review of both business-as-usual, global water scenarios (some of which are summarized in Chap. 2), a set of more positive scenarios, and second, a broader discussion of new approaches to thinking about water futures.

Gulbenkian Think Tank on Water and the Future of Humanity, *Water and the Future of Humanity: Revisiting Water Security*, DOI 10.1007/978-3-319-01457-9_8, © Calouste Gulbenkian Foundation 2014

We identify more positive alternatives to the business-as-usual scenarios described in Chap. 2. We discuss the development and use of "backcasting" approaches as another way to produce scenarios that more directly explore concepts, policies, and tools for sustainable water management. Finally, we present the concept of the "soft path for water" in order to help stimulate thinking—and perhaps specific strategies—for moving to a desired water future.

8.2   History of Water Scenario Development and Use

The future is, of course, largely unknowable. Scenarios should be treated as "stories," or possible futures, with the understanding that choices we make today determine which path we end up following and which future we move toward. But consciously or not, humans have always thought about possible futures, explored alternative possibilities, and tried to identify risks and benefits associated with societal choices. This has led to a growing interest in scenarios, forecasting, and "future" studies (see, e.g., Schwartz 1991). In the water sector, expectations about future water demands and supplies have implications for national budgets and financial expenditures, construction planning, and ultimately human and ecological well-being. Failing to make timely decisions and investments can have consequences for human health and economic stability.

There is a long history of the use of scenarios to project future needs for water. The time needed to develop major water facilities, and the subsequently long lifetimes of those facilities, require planners to take a relatively long view into the future. But what will future water supplies and demands be? How can they be estimated or evaluated, given all the uncertainties involved in looking into the future? How can concerns about sustainability be factored into long-term planning?

Many business-as-usual global projections and estimates of future freshwater demands have been produced over the past half-century, some extending out as much as half a century or more into the future. A number of these are described in Chap. 2. These projections have invariably turned out to be flawed—often substantially wrong. During the last 4 decades of the twentieth century, water resources planning focused on using or making simple projections of populations, *per capita* water demand, agricultural production, levels of economic productivity, and similar economic or demographic drivers. These drivers were used to estimate future water demands and then to evaluate the kinds of systems or structures that could be implemented or built to satisfy those demands. As a result, traditional water planning tended to project future water demands as variants or extensions of current trends, independent of any analysis of specific water needs or any assessment of approaches for modifying demands, i.e., demand management. Often these projections were done independently of estimates of actual regional water availability and often, the results were (in hindsight) completely unreasonable and disconnected from the realities of watersheds and the hydrological cycle.

Comparing among and across water scenarios is difficult. All scenarios differ in assumptions and the ways in which various drivers are used. Planners might assume,

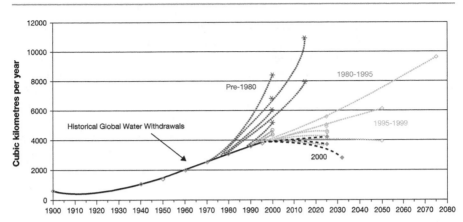

Fig. 8.1 Actual water withdrawals and scenario projections over time. *Red lines* show projections made prior to 1980. *Blue* and *green lines* show projections made between 1980 and 1999. The earlier scenarios tend to be based on business-as-usual assumptions; more recent ones typically include a broader range of assumptions of positive actions. *Black dotted lines* show three more recent scenarios that are discussed later in this chapter: Cosgrove and Rijsberman (2000), Rosegrant et al. (2002), and GEO-3 (UNEP/RIVM 2004)

for example, that international human rights policy results in greater commitments to domestic access to piped water, and such commitments can be quantified. Another approach might try to estimate environmental flow requirements in selected major river basins and deduct those requirements from estimates of water available for use by humans before allocating the remaining water to various users. Some scenarios go into more depth in the area of agriculture—the dominant consumptive use of water in most regions. Others pursue greater detail on the industrial or domestic sectors. Case studies and regional assessments play a large role in many scenarios. An important determinant in assessing the quality of any scenario is how transparent the creators have been with respect to the basic assumptions and data manipulations used.

Despite differences in approaches, all of the early global water projections estimated far greater demands for water than actually materialized, many by a substantial margin. This suggests that the traditional methods used by water-scenario developers miss important real-world dynamics (Gleick 2000a). Figure 8.1 shows a wide range of early water projections for various points during the twenty-first century along with an estimate of actual water use up to 2000. Most early projections greatly overestimated future water demands by assuming that use would continue to grow at, or even above, historical exponential growth rates. Actual global water withdrawals at the turn of the millennium were only around half of what they were expected to be just a few years earlier. Only in the last few years have newer projections included slower rates of growth or even the possibility of reductions in total water demands. We do not provide a comprehensive review of these traditional scenarios, but more information can be found in Gleick (2000a) and details on several are provided in Chap. 2.

For most water scenarios, data and assumptions are derived from a relatively small selection of global water resources studies and datasets to estimate current water use by region and sector, and, until recently, from a limited number of analytical models (e.g., WaterGAP). Baseline years vary, depending on when the scenario was published. Most recent ones typically use a starting point of 1990, 1995, or 2000. Scenario endpoints vary widely, with many projecting water use to 2025 or 2030, and a few looking further toward 2050 or even 2075.

Most early projections use variants on the same business-as-usual approach. The future water use is based on population and economic projections, simple linear assumptions of industrial, commercial, and residential water use intensity (e.g., water per unit population or income). Basic estimates of future crop production are a function of irrigated area and crop yield (see Gleick 2000a). Early scenarios are typically single projections with no or few variants. Most ignore important components of societal or environmental water use, such as water requirements for in-stream ecological needs, navigation, hydropower production, recreation, and other important uses. Most also ignore the "green" water component, focusing solely on "blue" water withdrawals and use (Falkenmark and Rockström 2006).

Scenarios also differ in their approach to scenario design, model/simulation construction, phases of iterative development, format/presentation of findings, and identification of solutions. Traditional scenarios begin with a background/current assessment of water resources that is projected forward in a business-as-usual future scenario. Alternate scenarios typically derive from models that manipulate the same baseline data used in the business-as-usual to produce modest variants.

A range of drivers is usually found underlying these scenarios, spanning the following major categories:

• Large-scale economic and demographic drivers (population change and gross domestic product)
• Technology and infrastructure (efficient irrigation or industrial water use technology)
• Climate and hydrology (temperature and precipitation, sometimes factoring in for climate change)
• Policy and governance (management institutions and water pricing)
• Environment (dedicated in stream flows)

Agriculture and food production often stands as its own category in these scenarios because agriculture accounts for the majority of human consumptive use of water. Some scenarios stress social/cultural drivers such as the impact of international human rights policy or participatory governance of water resources, although these drivers do not readily lend themselves to quantification. Water quality is almost never a factor in any of the scenarios.

In the last few years, new approaches to making such projections have been discussed and tested. New efforts are taking advantage of improvements in computing, the availability of better water data, and new concepts of scenario development. But such newer efforts are rare and still constrained by a range of challenges.

8.3   Data Constraints

One of the most serious challenges to improving efforts to develop and use water scenarios is the poor quality, availability, and regional resolution of water data. This is especially so for data on historical water uses at high sectoral and regional resolution. These limitations have been understood for some time, but most remain unresolved. For example, there are almost no global or national estimates of total water withdrawals and consumption. Exceptions are those compiled by Shiklomanov (1998) and his colleagues at the State Hydrological Institute of St. Petersburg and those that originate with estimates presented in the UN Food and Agricultural Organization (FAO) AQUASTAT database. Among the important data problems are the following:

- *Serious gaps in regional-scale hydrological data exist.* Many regions, even in more developed countries, suffer from lack of reliable long-term records for fundamental components of the hydrological cycle, including precipitation, evaporation, runoff, and most groundwater information. In recent years, even existing monitoring systems are being threatened with closure due to lack of funding and commitment to collect data (Fig. 8.2). Some gaps are being filled by satellite and other remote sensing systems. However, these systems are not yet comprehensive in coverage and national budgetary problems also threaten these systems.
- *Certain types of water use data are not collected or reliable.* Basic information on the current state of human use is needed for all scenarios of future water needs.

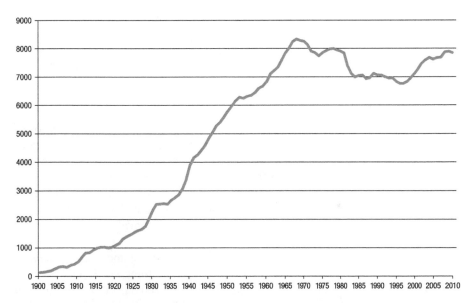

Fig. 8.2   The number of active United States Geological Survey (USGS) stream gauges from 1900 to 2010. *Source*: USGS (2012)

Yet far less data are collected on water use than on water supply and availability, or the hydrological cycle. Water use is often not measured directly; rather it is often modeled or estimated (as for much data in the AQUASTAT database). Industrial and commercial water uses are reported infrequently or not at all. Agricultural water use data are even more uneven and unreliable. Groundwater withdrawals are rarely measured, regulated, or compared with natural recharge rates. Information on changing water use patterns over time is often not available, making analysis of trends difficult. Further complicating this problem are the different definitions of water "use," including consumptive, nonconsumptive, withdrawals, reuse, and more (Gleick et al. 2011).

- *Water-quality data are especially limited or unreliable.* While there are gaps in many different kinds of water data, water-quality data are particularly difficult to collect and access. Water quality varies substantially over space and time, requiring systematic and long-term data collection. The enormous range of water-quality contaminants and the difficult and expensive nature of collecting and analyzing contaminant types and levels further complicate these efforts (Palaniappan et al. 2010). Finally, even when efforts are made to collect and collate water-quality data, such as the UN Global Environment Monitoring System Water Programme's (GEMS) efforts, they are often underfunded or constrained by requests that data not be publicly disseminated.

- *Some countries or regions still restrict access to water data.* Some countries and regions refuse to share water-related data with neighbors or even their own scientists despite the call in the 1997 Convention on the Law of the Non-Navigational Uses of International Watercourses for open sharing of data. Article 9 of that Convention, which has still not come into force at the time of this publication, states:

  ...States shall on a regular basis exchange readily available data and information on the condition of the watercourse, in particular that of a hydrological, meteorological, hydrogeological and ecological nature and related to the water quality as well as related forecasts... [States] shall employ their best efforts to collect and, where appropriate, to process data and information in a manner which facilitates its utilization by the other watercourse States to which it is communicated.

- *Some water uses or needs are unquantified or unquantifiable.* Most scenarios of future water use still fail to include water needs for ecosystems or other non-standard demands for water, such as, reservoir evaporation and seepage or evaporative cooling for agriculture. Information on water reuse (including so-called *gray* water) is also rarely available. Many of these uses have never been quantified, which makes them hard to include and analyze. Nevertheless, comprehensive water planning will ultimately require improved information on these water characteristics and requirements.

As a result of these data constraints, improvements in models or scenario sophistication do not necessarily lead to more accurate forecasts. Even "perfect" models (if they could be created) provided with imperfect data will be flawed. As a result, scenarios should always be treated as "stories"—as possible futures—rather than as deterministic endpoints. As the British statistician George E. P. Box noted, "Essentially, all models are wrong, but some are useful." Many critics of models focus on the first part of this observation and ignore the second.

8.4   Positive Water Scenarios

As described in Chap. 2 and elsewhere, a wide range of classic, business-as-usual type water scenarios have been produced over the past half-century. These scenarios represent a range of the methods, assumptions, and drivers used in scenario analysis. They offer a diversity of outcomes with the limitations described earlier. Despite these limitations, interest in developing water scenarios remains high as a way to explore future paths, water policies, and larger environmental or sustainability strategies. But in addition to these traditional business-as-usual projections of the future, many analysts attempt to develop alternative scenarios that provide a vision of a more "positive" or desirable future. These scenarios are developed using the same methods but altering the underlying drivers and assumptions to some degree.

This section describes some key factors that can be manipulated to, at least somewhat, change the expected trajectory of future water demands toward a more "sustainable" or positive future. Projections have begun to include reassessments of actual water needed to produce mixes of goods and services (using water "footprinting" methods), the technical potential for water use efficiency improvements, alternative dietary or caloric requirements, modified cropping patterns and types, climatic variations, and ecosystem water needs. None of these assessments have used a strict "backcasting" method. However, the model formulation and iteration process described in some models yields a similar effect—where expected future outcomes (such as technology or policy changes) determine how input data are manipulated using major drivers (population change, technology, and the balance of water available to humans and/or the environment). Except on a limited regional basis, no quantitative endpoint water target (and rarely even qualitative targets), such as, an imposed water usage cap or limit based on a vision of sustainable water allocation and use, has yet been applied. For example, it is rare that constraints on our capacity to extract water with available technology have been used to identify an upper limit of water use. Recently, the concept of "peak water," including peak renewable, nonrenewable, and ecological water, has been introduced. This too has not yet been applied to scenario efforts (Gleick and Palaniappan 2010).

The simplest way to produce more positive scenarios is simply to modify the assumptions of the major drivers. As noted earlier, the most common major drivers for projections of water demand are population and macroeconomic indicators such as GDP. When the major determinants of water use are estimates of water "intensity" (defined as water use per person or per unit of economic production), reductions in intensity lead to reductions in water demands. Almost all of the most well known of the scenario projects of the 1990s and 2000s developed positive water scenarios using modifications of these factors (e.g., Alcamo et al. 1997; Raskin et al. 1997, 1998; Seckler et al. 1998).

Another approach to producing positive water scenarios is to alter the water use coefficients related to the production of key goods and services, especially industrial or agricultural output per unit water. In the real world, such improvements come about through changes in technology. For example, shifts from flood irrigation to precision irrigation, or industrial processes that reduce the water requirements for manufacturing. In general water scenarios, however, more simple changes in assumptions are used to mimic these productivity improvements, permitting greater production with less water. Most "positive" scenarios apply assumptions of an improvement in efficiency to represent these real-world improvements. Alcamo et al. (1997), for example, included some generic "efficiency improvements" and "conservation targets."

Improvements in computer technology and analytical methods have made it possible to begin to integrate specific assumptions about water policies and technologies to produce alternative pictures of water demands. Seckler et al. (1998) included a detailed assessment of the irrigation sector with different assumptions about the productivity of agricultural water use. They looked at variables such as the effectiveness in use of irrigation water, changes in irrigated area and cropping patterns, and changing reference evapotranspiration rates for different countries and seasons. In addition to the standard business-as-usual scenario, a more positive scenario was developed. This assumed a high degree of effectiveness in the use of irrigation water with an exploration of changes in irrigated area, withdrawals of water per hectare of irrigated area, and reference evapotranspiration rates for different countries and seasons. Alcamo et al. (2007), e.g., included the effects of income, electricity production, water use efficiency, and other driving forces on water stress.

In 2007, the Comprehensive Assessment of Water Management in Agriculture (CAWMA 2007) explored scenarios of agricultural water use productivity improvements. That report argued that improving rainfed agricultural production, increasing irrigated agricultural productivity, and enhancing trade could hold irrigation withdrawals at approximately 3,000 km^3 per year and total water use in irrigated and rainfed agriculture to just over 8,000 km^3 per year. This was far below traditional business-as-usual estimates. However, to achieve this outcome, the authors emphasize that major gains in productivity (crop per drop) are needed, particularly in the developing world. Similarly, significant reforms to water governance, water management practices, and trade policies would be required. Special efforts would be needed to maintain and improve ecosystem services.

Another approach for developing positive future scenarios is to apply a range of policy options and choices that are then assumed to directly affect total water requirements. Examples are using water pricing combined with income elasticity assumptions to influence future demands, or altering social values and lifestyles, such as diet, to change water needs. The World Water Council (WWC) produced a book, coauthored by William Cosgrove and Frank Rijsberman, addressing future water scenarios (Cosgrove and Rijsberman 2000). An extended discussion of the details of this effort can also be found in Gallopin and Rijsberman (2000). The authors offered three scenarios for 2025, including a business-as-usual scenario (from a base year of 1995) and two alternative scenarios. One of the alternative scenarios was an "Economics, Technology and Private Sector" scenario (TEC). This relied on policies favored by those who prefer market solutions, the involvement of the private sector and technological solutions—with most efforts focused at national or basin scales. The other was a "Values and Lifestyles" scenario (VAL). This included "a revival of human values, strengthened international cooperation, heavy emphasis on education, international mechanisms, international rules, increased solidarity and changes in lifestyles and behaviour." The VAL scenario explored how increases in water demands can be offset by efficiency improvements and the saturation of water demands in industry and households. Among the key "levers" evaluated are the expansion of irrigated agriculture, the productivity of water use, increasing storage, reforms of water-resource management institutions, increasing cooperation in international basins, valuing ecosystem functions, and supporting innovation. The VAL scenario projected that the amount of irrigated land will ultimately stabilize and that water use efficiency for irrigation will improve, leading to leveling off and perhaps even a decline in overall water withdrawals.

The United Nations Environment Programme (UNEP) and the National Institute for Public Health and the Environment (RIVM) conducted a similar exercise in the Netherlands in 2000. That report assessed the impact of climate change and socio-economic changes on cropland, built environments, hunger, ecosystems, and water resources in Europe and the world between 2002 and 2032 (UNEP/RIVM 2004). One chapter was devoted to "water stress" with two positive scenarios that yield improved environmental outcomes. The scenarios were split into global, sub-global regions, and country assessments with a baseline of 2002 (using data from either 1995 or 2000, modified) and an ending date of 2032. The "positive vision" scenarios were labeled "Sustainability First" and "Policy First," with the major drivers being social changes and policy changes, respectively. These scenarios were largely developed using qualitative assumptions, backed with quantitative assessments (Cosgrove and Rijsberman 2000). In that study, socioeconomic drivers were used to address environmental impacts. Policy actions and behavioral changes were applied to accelerate the transition to slower growth. Subsidies and regulation (e.g., pollution taxes) were applied both to improve technology and efficiency and to reduce demand.

An approach that combines many of these methods is that of Rosegrant et al. (2002). They analyzed long-term (30-year) projections of domestic, industrial, live-stock, and irrigation water demand and supply. They looked at nearly 70 individual or aggregated river basins at a global scale and incorporated various factors not found in most traditional water scenarios. These factors included seasonal and inter-annual climate variability and especially, a "positive" scenario that incorporates sustainability and efficiency measures, environmental water allocations, complete provision of piped water to all urban households, improvements in water use efficiency, and increased per capita domestic water consumption while maintaining food production. On the institutional side, their positive scenario also looked at institutional changes, e.g., market-oriented reforms in the water sector, "more comprehensive and coordinated government action," and greater investments in both infrastructure and water use efficiency and productivity improvements.

8.5   Results for Positive Water Scenarios

As might be expected, the "positive" scenarios all result in less total water demand than the related business-as-usual scenarios. In some, total water demand actual decreases over time from current use, though most still show increases from

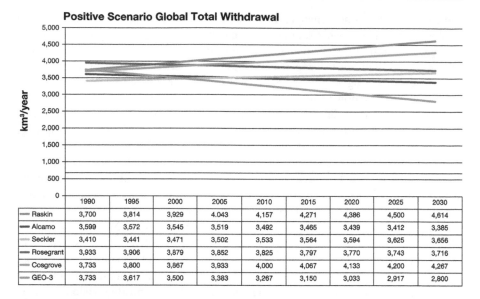

Fig. 8.3 Comparison of total global water withdrawal in selected "positive vision" scenarios. Baseline and endpoint years differ slightly among the six scenarios above, although most use 1995 as their baseline and 2025 as their endpoint. Interim years for which scenarios do not provide explicit estimates are derived here through linear interpolation. Note that in contrast to the business-as-usual scenarios, there is a wide spread of endpoints, with the GEO-3 scenario leading to a significant decrease in absolute water withdrawals (UNEP/RIVM 2004), while the Raskin et al. (1997) and Cosgrove and Rijsberman (2000) scenarios show increases. Most of the others show modest decreases in total withdrawals over time

baseline levels. Figure 8.3 shows total global water withdrawals over time for some of the positive vision scenarios.

While the greatest absolute differences among the positive vision scenarios are apparent in the agricultural sectors, the greatest percentage differences occur in the industrial and domestic sectors. For example, industrial water demands for a positive vision in 2030 range dramatically from over 800 km^3/year for Rosegrant et al. (2002) and Cosgrove and Rijsberman (2000) to around 200 km^3/year for Alcamo et al. (1997)—a fourfold difference. The differences are far smaller for domestic withdrawals—twofold between the GEO-3 (UNEP/RIVM 2004) and the Cosgrove and Rijsberman (2000) scenarios. In contrast, for agriculture, the differences are around 1,900–2,600 km^3/year for Alcamo et al. (1997) and GEO-3 (UNEP/RIVM 2004) on the low end; and Cosgrove and Rijsberman (2000) on the high end—a difference of around 30 % from the low estimate (Fig. 8.4).

These "positive vision" scenarios suggest that it is possible to avoid the negative aspects of traditional projections. However, the methods used to develop them, and the kinds of policies and practices incorporated into the models, are little different from current approaches. Yet we also know that there are serious economic, environmental, and social limitations and liabilities to current water policies—as described in many of the preceding chapters. As a result, another method of evaluating water futures has been developed over time—*backcasting*.

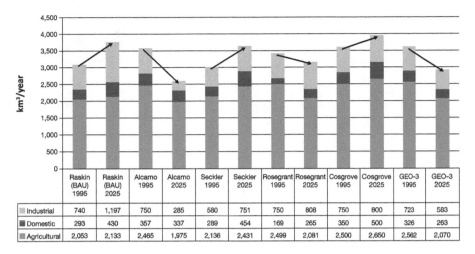

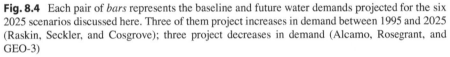

	Raskin (BAU) 1995	Raskin (BAU) 2025	Alcamo 1995	Alcamo 2025	Seckler 1995	Seckler 2025	Rosegrant 1995	Rosegrant 2025	Cosgrove 1995	Cosgrove 2025	GEO-3 1995	GEO-3 2025
Industrial	740	1,197	750	285	580	751	750	808	750	800	723	583
Domestic	293	430	357	337	289	454	169	265	350	500	326	263
Agricultural	2,053	2,133	2,465	1,975	2,136	2,431	2,499	2,081	2,500	2,650	2,562	2,070

Fig. 8.4 Each pair of *bars* represents the baseline and future water demands projected for the six 2025 scenarios discussed here. Three of them project increases in demand between 1995 and 2025 (Raskin, Seckler, and Cosgrove); three project decreases in demand (Alcamo, Rosegrant, and GEO-3)

8.6   Backcasting Approaches

All future projections of water supply and demand depend on assumptions about fundamental drivers. Such assumptions include population size, growth rates, and distribution, and economic factors, climatic conditions, technology development and penetration rates, and institutional policies. As already mentioned, most traditional water scenarios have relied upon a very limited number of influential drivers, especially population size and assumptions about economic productivity indicators such as water per unit GNP. These drivers are largely independent of water policy decisions and assume that past relationships between the drivers and water use remain unchanged in the future. As a result, traditional business-as-usual scenarios such as those discussed in Chap. 2 do little more than lead to unsustainable futures with the same kinds of water problems we already see today.

In the past two decades, planners and policy makers have begun to look at alternative approaches to developing scenarios. In particular, the growing interest in "sustainable development" and integrated water resources planning has spurred the use of backcasting, where desirable futures or visions are described and paths back from that future to the present are explored (see Fig. 8.5) (Holmberg and Robert 2000; Quist 2007; Phdungsilp 2011).

Among the standards steps in a backcasting exercise are:
- Analysis and definition of the "problem"
- Development of a normative, often desirable future vision
- Creation of a process to determine what is necessary to reach that vision

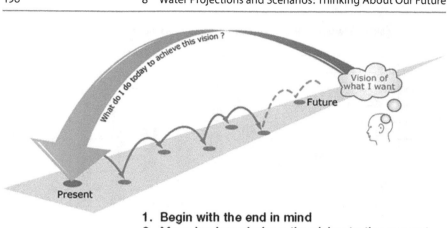

1. **Begin with the end in mind**
2. **Move backwards from the vision to the present**
3. **Move step by step towards the vision**

Fig. 8.5 Typical "backcasting" approach. *Source*: The Natural Step (2012)

- Elaboration and analysis of an "action agenda"
- Implementation and review of effectiveness of actions
- Evaluate whether by these actions the desirable future is likely to be achieved
- If necessary adjust either or both the vision and the action agenda until the desired outcome is achieved

Visions developed during backcasting projects are no more or less likely than business-as-usual scenarios to occur. But backcasting offers a different kind of tool for work in the field of sustainable development, including both water and energy planning. In particular, it provides a way of identifying successful and sustainable futures and permits exploration of the question "What is necessary to reach any of those futures?"

One of the key advantages of backcasting is to move away from the focus on non-water-related drivers. Instead the drivers used are more directly influenced by explicit water policy decisions. This includes investment in water use technology, pricing structures for water utilities and users, strategies to alter investment in irrigation systems or shifting to other kinds of water supply systems, integration of climatic factors that directly influence water supply and demand, or adoption of dietary choices combined with varying assumptions about cropping patterns and rainfed production. By expanding the number and kinds of drivers, water planners can evaluate strategies to influence or change their effect.

8.7   Backcasting Experience in Water Planning

One of the greatest weaknesses of traditional forecasting based on extrapolation of current trends is the difficulty of generating solutions that require shifting strategies and breaking trends. In other words, it is hard to reach a future you cannot envision or to change paths without changing strategies. Backcasting approaches can help remove these barriers. Weaver et al. (2000) describe backcasting as a tool to evaluate desirable futures and for helping provide a systems perspective. They describe a series of steps that lead to the development of a long-term approach to assess how specific needs might be met in the future and backward analysis to lay out strategies (policies, technologies, and tools) for sustainably reaching that future. This includes both short- and long-term actions as well as approaches to implementation with appropriate stakeholders. Ideally, backcasts should include socioeconomic, environmental, technological, and policy dimensions. These should be used in a way that helps avoid impossible futures, implausible economic or social outcomes, and the reliance on unrealistic technological or economic developments. Such unrealistic visions may be both hard to avoid and hard to identify in advance.

Backcasting was first explored for energy analysis, urban planning, and transportation, especially in Europe and in some particular industrial sectors, like energy (Lovins 1976, 1977; Robinson 1982). Within the energy field, one of the earliest efforts to develop real alternatives to traditional scenarios was the work of Amory Lovins, who used it to develop far different electricity supply and demand requirements than previously seen (Lovins 1977).

There have been only limited efforts to apply the approach in the water sector. Water backcasts, like those for the energy sector, define particular objectives and goals, identify constraints to reaching those goals, develop positive scenarios based on both the objectives and the constraints, and evaluate strategies for reaching the target (Holmberg and Robert 2000). In the water sector, some of the earliest efforts include the work of the Pacific Institute in positive backcasts for California water resources (Gleick et al. 1995) and "soft path for water" work (Gleick 2002, 2009b; Brooks and Brandes 2005; Brooks et al. 2009).

In developing a sustainable water system that is realistic and achievable, it is important to retain the best of what already exists. Scenarios that require completely rebuilding or reinventing either infrastructure or institutions (at least in the short run) are unrealistic. At the same time, scenarios that include reasonable modifications to infrastructure, options presented to policy makers, financial and economic approaches, and institutional organizations can help planners move away from the business-as-usual paths (especially those that clearly have undesirable outcomes).

For water planners, the first step in backcasting requires that standards or clear objectives in the form of sustainability principles be identified, such as those developed for the "soft path for water" approach (Gleick 2002). If these principles are properly defined and met, a vision of future, sustainable water policies can be prepared. In 1995, for example, a positive backcast for California water resources was developed that defined sustainable water systems as those having the following characteristics (Gleick et al. 1995, p. 100):

1. A minimum water requirement will be guaranteed to all humans to maintain human health.
2. Sufficient water will be guaranteed to restore and maintain the health of ecosystems. Specific amounts will vary depending on climatic and other conditions. Setting these amounts will require flexible and dynamic management.

3. Data on water resources availability, use, and quality will be collected and made accessible to all parties.
4. Water quality will be maintained to meet certain minimum standards. These standards will vary depending on location and how the water is to be used.
5. Human actions will not impair the long-term renewability of freshwater stocks and flows.
6. Institutional mechanisms will be set up to prevent and resolve conflicts over water.
7. Water planning and decision making will be democratic, ensuring representation of all affected parties and fostering direct participation of affected interests.

Backcasts would also have to address all forms of water resources. This includes "blue," "green," and "gray" (Falkenmark and Rockström 2006), and integrating the risks and vulnerabilities of water resources (both supply and demand) to global climatic changes, which are already affecting water resources in many regions.

While developing future "sustainable visions" for water may be difficult, identifying strategies and paths to reach those visions are often considerably harder. Natural and human systems are complex and nonlinear. Institutional systems are slow to change, even with intent and perfect information, both of which are often lacking. The advantage, however, of developing backcasts and positive visions is that they help to develop common objectives and to work toward common positive changes in societal perceptions and actions around freshwater.

Box 8.1 A Global Water Backcast

The Pacific Institute (Gleick 1997) prepared one of the earliest water "backcasts." It developed a disaggregated "end-use" approach instead of traditional supply/demand projections and generated a "Vision" scenario for the year 2025. Model assumptions included future water use by region and sector (domestic, industrial, and agricultural) under a set of explicit sustainability criteria and limits. In the Vision scenario, total domestic water use in 2025 was estimated using two assumptions. First, the world's entire population has access to a "basic water requirement" of 50 L per person per day to meet basic needs (Gleick 1996). Second, regions using over that amount in 1990 implement water-productivity improvements to reduce *per capita* domestic water use toward the level presently used in the more efficient nations of Western Europe—around 300 L per person per day. The net result is that total domestic water needs in 2025 would not be substantially different to 1990 estimates. This would be despite a large increase in population. The distribution of that water use would also be far more equitable than today's distribution—a key criterion for sustainability as defined by the "soft path" approach (Gleick 2002).

Agricultural water use projections were also based on end-use assumptions—in this case specific human dietary needs in each region and the water requirements to produce calories of specific food types. The Vision scenario

(continued)

Box 8.1 (continued)

assumes that *per capita* meat consumption in Europe and North America drops, while overall calorie consumption grows in the developing world toward a minimum health standard. In particular, all regions are assumed to reach a minimum of 2,500 cal per person per day by the year 2025. Those regions currently consuming over 3,000 cal per person per day move toward dietary changes that reduce consumption toward 2,500 cal per person per day. There is no discussion in the study of the specific policies (regulation, education, pricing, etc.) required to achieve this dietary shift. A more comprehensive backcast effort would explore such policies (see, e.g., Chap. 4).

The total water needed to grow these diets is substantially below typical estimates as a result of a reduction in water-intensive components, particularly meats. A North American diet, e.g., that requires over 5,000 L per person per day to grow today would be reduced to 3,500 L per person per day. This is still a more water-intensive diet than average but it is also a considerable water saving from business-as-usual. Similar reductions were developed for each geographic region along with assumptions about improvements in irrigation efficiency and cropping intensities. With these assumptions, the Vision scenario projected that overall irrigation requirements would still rise during 1990–2025, although far less than in conventional development scenarios.

Future industrial and commercial water demands could also look significantly different than under traditional futures. By 2025 in the Vision scenario, total industrial water withdrawals remain virtually the same as 1990 levels. However, there are far higher levels of productivity and a more equitable distribution in *per capita* industrial water use than today. This was achieved by implementing policies to improve the industrial productivity of water use, expand use of recycled water, and shift industrial structure and efficiency. *Per capita* industrial water use in almost all developed regions would drop—most dramatically in Europe and North America. This would increase in Asia, Africa, and Latin America on both a *per capita* and absolute level, through technology improvements driven by price or regulatory strategies. The opportunity to bypass certain styles of development would permit many nations to move directly to industries and energy systems that consume less water.

8.8   The Concept of "Water Wedges"

Another component of backcasting is the ability to integrate information about both policy and technology approaches and to evaluate the potential of different strategies to influence the path over time. Each of these strategies can be considered a "wedge," and the combination of wedges makes up (in theory) an overall,

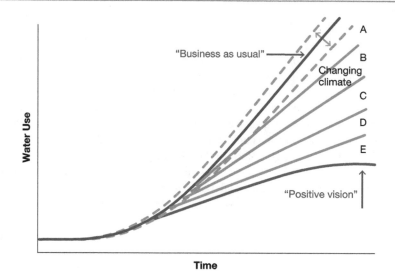

Fig. 8.6 A business-as-usual and a "positive vision" scenario, with hypothetical "wedges" (*A–E*) that represent policy or technology options for either reducing water demand or increasing supply. Each wedge can be the subject of detailed assessment and analysis

integrated water policy. Figure 8.6 shows a hypothetical business-as-usual water projection over time, with a typical exponential increase in expected demand. Also shown in this figure is a much lower demand projection, perhaps developed through a backcasting approach, or a standard forecasting approach with substantially different assumptions. In either case, the difference between the two endpoints is the total reduction in water use resulting from the application of water policies. Each assumption accounts for a "wedge" of potential water demand reduction and a deviation from the business-as-usual path. This "wedge" concept is modified from the climate/energy work of Pacala and Socolow (2004). It offers insights into the effectiveness of different approaches in the context of addressing global climate challenges and the need to reduce greenhouse gas emissions. In the water context, such an approach can lead to both more detailed analysis of strategies for demand management (or supply augmentation) and specific water policy recommendations.

Such policies can include (and are presented in Fig. 8.6):

- Technical choices to augment water supply (such as pipelines/aqueducts, use of recycled water, desalination, and local rainwater harvesting) or modify demand (such as efficient appliances, metering/measurement, and crop genetics)
- Economic strategies to augment supply (such as subsidies and tax credits/incentives) or modify demand (such as pricing, rate design, and subsidies)
- Educational strategies to modify demand (such as information on dietary choices, or social suasion on water use behavior, or information sharing among neighboring farmers)
- Regulatory strategies to modify demand (such as drought restrictions, appliance efficiency standards, or land use policies)

8.9   Soft Water Paths

Traditional water development strategies—successful as they have been in some ways—are increasingly recognized as inadequate for addressing modern water problems. We must now find new approaches and a new path, but the process of defining, developing, and implementing such alternatives has yet to be completed. Despite evidence about the limitations of old approaches, presented in earlier chapters, most planners, utilities, engineers, and designers in the water sector continue to focus on traditional physical infrastructure (such as dams and centralized water treatment and distribution systems) and management (such as centralized public or private water or agriculture agencies and state planning ministries)—the "hard path."

The traditional hard path approach produced enormous benefits for billions of people. These included clean and reliable water supplies, irrigation water in arid lands, and improved human health. However, many traditional projects that were designed and built either failed to deliver their promised benefits or caused a wide range of ecologically damaging, socially disruptive, and economically costly impacts (Gleick 2002).

There is an alternative. In the past decade or so there has been a growing interest in defining a new approach—the "soft path." This capitalizes on the advantages of traditional infrastructure and planning, but complements it with a wider range of tools and strategies that are more environmentally, socially, economically, and politically effective. As defined by Gleick (2002), Wolff and Gleick (2002), and others (see Brooks et al. 2009), the soft path continues to rely on carefully planned and managed centralized infrastructure but increasingly complements it with small-scale decentralized facilities and management. The soft path for water works to improve the water use productivity rather than simply expanding traditional sources of supply. It delivers water services and qualities matched to users' needs, rather than just delivering quantities of water. It applies economic tools, such as markets and pricing, as a way to encourage efficient resource use, equitable distribution and allocation, and sustainable operation and management. And it includes local communities in democratic decisions about long-term water planning and management. As Lovins noted for the energy industry, the industrial dynamics of a soft path approach are very different, the technical risks are smaller, and the dollars risked far fewer than those of the hard path (Lovins 1977).

One of the major strategies of the soft path is to rethink how and why we use water. Hard path planners incorrectly assume that demand (and hence supply) of water will continue to grow exponentially with population and economies. They equate the idea of using less water with a loss of well-being. This is a fallacy. Soft path planners understand that water "use" is not an end. It is a means to a variety of ends, such as producing goods and services. Our ultimate goal is not to use water but rather to satisfy desires for food, fiber, industrial, and commercial goods, and to satisfy domestic needs for cooking, cleaning, and waste disposal, and more. Society should not care how much water is used—or even whether water is used at all—as

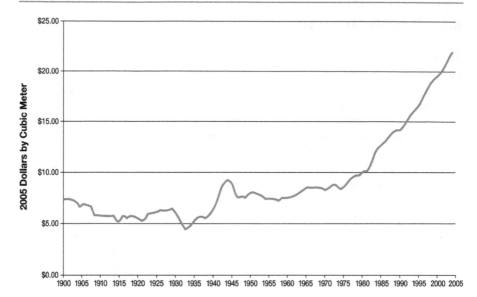

Fig. 8.7 Water productivity for the economy of the United States from 1900 to 2005 measured in USD of GNP per m³ of water used. *Source*: Data taken from Johnston and Williamson (2005) and Kenny et al. (2009)

long as these goods and services are produced in productive, cost-effective, and socially acceptable ways.

Approached in this manner, appropriate measures of well-being are "productivity" or "intensity" measures, such as social and individual well-being or the production of goods and services, per unit water used, as opposed to simple measures of total water use. The units of output can be physical (e.g., kilograms of wheat per unit of water) or economic (e.g., the dollar value of the good or service produced per unit of water). Figure 8.7 shows water productivity for the economy of the United States from 1900 to 2005 measured in USD of GNP per unit of water used. This figure shows that productivity was relatively constant until the 1970s. Then a combination of factors (such as rising environmental awareness, advances in technology, and the shift toward a service economy) caused water productivity to rise steadily. It is now over double the value in the 1970s and an indicator of a shift toward the soft path (Gleick 2002, 2009b).

The soft path offers great potential, though much work remains to be done to both identify and then achieve the characteristics of the soft path (Brooks et al. 2009). Among those characteristics are:

1. The soft path redirects water institutions at various scales (government agencies, private companies, and individuals) to satisfy water-related needs rather than merely supply water.
2. The soft path acknowledges that different water needs can be satisfied with waters of different qualities and that we match the quality needed with the quality available. Thus, high-quality, more costly (e.g., potable) water can be reserved for uses that require it, while lower-quality waters (such as storm water runoff,

gray water, and reclaimed wastewater) can be used to satisfy a wide range of other water needs.

3. The soft path requires water planners and managers to communicate more clearly and closely with water users and to engage community groups in water management. In contrast, the hard path has traditionally been governed by engineering approaches that satisfy large-scale generic needs with little or no community consultation.

4. The soft path recognizes that the health of aquatic ecosystems and the societal activities that depend on them are central to water policy and management. In contrast, the hard path assumes that water left in a river or lake or aquifer is not being used productively and thus has no value.

5. The soft path recognizes the complexity of water economics and the importance of proper full-cost analysis and pricing, equity, and investments in decentralized solutions. The hard path looks at projects, revenues, and economies of scale. The soft path also considers the economies of "scope," when combined strategies would permit improved efficiencies (Wolff and Gleick 2002). For example, water providers can often reduce the total cost of services by accounting for the interactions that separate agencies (such as energy or waste disposal utilities) cannot account for alone.

This soft approach to water management can also encompass and expand upon what is known as Integrated Water Resources Management (IWRM). The rationale for the IWRM approach has now been accepted internationally as a way forward for efficient, equitable, and sustainable development and management of the world's limited water resources and for coping with conflicting demands. The Global Water Partnership states: "IWRM is a process which promotes the coordinated development and management of water, land and related resources, in order to maximize the resultant economic and social welfare in an equitable manner without compromising the sustainability of vital ecosystems." In fact, this more inclusive and comprehensive approach to water resources planning and management is what most practitioners are striving to do today, given their new financial, physical, legal, environmental, and institutional constraints (WWAP 2009).

8.10   Strategies for Moving to a Positive Future

Explorations of water "futures" of any kind, whether they are business-as-usual scenarios, positive water scenarios, backcasts or soft path "visions," depend on implicit or explicit assumptions about technologies, economic and policy approaches, social and political structures, and demographics. Successful water management inevitably requires combinations of regulations, economic incentives, technological changes, and public education. Considerable experience in every sector of the economy suggests that the most effective programs include combinations of all of these approaches.

As noted above, every one of these assumptions can be considered a "wedge" that influences either future water supply or demand, and it is appropriate to assess what assumptions and drivers create the most significant change in sector water use within individual scenarios. This approach can help identify what assumptions play the most important roles or can be influenced by different actors or institutions. There are two key kinds of assumptions about drivers: those that are water-independent and those focused on water policy.

8.10.1   Water-Independent Drivers and Strategies

Key drivers of water use in most water scenarios are demographic and economic variables. In these scenarios, domestic use is often a simple function of population and a minimum or estimated *per capita* consumption value, or calculated from GDP projections (and then subsequent assumptions about the infrastructure and efficiency with which GDP is produced in terms of water). There is little water managers can do about population or economic growth rates and as a result they mostly use official population projections unchanged. The range of *per capita* domestic water use is sometimes determined by the wealth of a country. More developed countries will have greater *per capita* consumption and less developed countries will have less. Across the board, the positive scenarios assumed that populations in less developed countries (expected to increase faster than populations in more developed countries) achieve a higher, or at least minimum, daily domestic allowance than at present. They also assume that *per capita* water consumption in more developed countries decreases with savings derived from conservation and efficient technologies.

8.10.2   Water-Focused Drivers and Strategies

More recently, however, there has been a growing effort to develop water scenarios that look explicitly at water-focused drivers rather than water-independent ones. These are strategies that can be implemented by water managers or water users directly. Hence, they can more directly influence water supply and demand. While there are many ways of describing or analyzing these strategies, we group them below in four categories: technological, regulatory, economic, and educational. A comprehensive water policy will offer a balance among these strategies rather than relying on one single approach, and the balance will, of course, vary depending on regional cultural, social, and political factors.

Technological tools include all of the approaches that permit us to collect, store, treat, and distribute water; use water more productively or manage it more effectively than in the past, treat and reuse water or convert previously unusable sources into usable ones. Typical technological tools used in scenario assessments include all approaches related to water use efficiency through changes in end-use appliances (e.g., toilets, irrigation systems, and cooling systems) or methods for expanding water supply (e.g., reservoirs/dams, water treatment and reuse plants, fog harvesting, and rainwater collection). Also included in technological options are tools to monitor and share water data, including both fundamental data on the hydrologic cycle, and information on how humans and natural ecosystems use water. This can include sophisticated remote sensing from satellites or ground-based sensors, or simple low-cost community reporting of water use and quality.

Regulatory tools include policies taken by governments to encourage water conservation and efficiency improvements, better water-quality management, and groundwater oversight. This also includes the setting of appliance efficiency standards, adopting landscape ordinances, and developing efficient building codes. Regulations play a special role in standardizing the rules and behaviors of diverse actors, such as when water-quality standards are put in place to reduce the "tragedy of the commons" associated with traditional water pollution risks.

Economic strategies can be highly effective in changing water paths and choices. Such strategies include the design of rate systems, marginal cost volumetric water prices, innovative insurance programs for farmers or landowners, rebates for water saving end-use devices or practices, low-cost loans or assistance in obtaining credit for capital investments by customers, and environmental fees or surcharges that allow compensation for those damaged by additional water withdrawals (e.g., fishermen). They also involve actual trading of water, water rights, or water use permits. Economic tools have been shown to be highly effective in modifying water supply and demand under certain circumstances. However, care must be taken to ensure that factors that fall outside of markets, or are difficult or impossible to quantify in economic terms (such as many ecosystems services), are not ignored.

Educational approaches include ensuring that information on options, costs, technologies, and regulations are fully available to water users. Educational strategies have long been used in the water supply, sanitation, and hygiene areas to teach communities the health benefits of water improvements. Governments also try to use these approaches to change behavior around water use during droughts and flood emergencies or to inform consumers about the water implications of their purchases and actions. Educational strategies can include advanced media and informational techniques to communicate with and persuade water users to behave in ways that are socially desirable (i.e., achieve social objectives appropriate to each locale). Smart choices will only be made when those choices are known and understood. This also requires that efforts to improve the flow of water-related information to both planners and end-users have been shown to be effective at changing the nature and magnitude of human impact on the water cycle.

8.11   Moving from Here to There: Where Do We Want to Be?

Natural and human-built water systems face substantial unsolved challenges in coming years. This is especially so since traditional solutions no longer seem effective, and increasingly because fundamental assumptions about stability in both climatic and political systems no longer seem valid. One response to these challenges requires doing sophisticated strategic planning. This needs to incorporate multiple factors, including past, present, and future conditions and drivers (such as population, economic growth and policies, technological changes, and societal preferences), alternative climatic conditions, long-term visions of the public and policy makers, and institutional capabilities and limitations. These items are not independent. Ideally, they would be addressed in a systems-oriented perspective. Futures studies, scenarios, and backcasting can offer insights not usually obtained from more traditional scenario efforts. These come with the belief that by doing so, it is possible to influence changes in current strategies and approaches and alter the long-term trajectory of investment strategies, policy decisions, and public priorities.

A wide range of water scenarios have been developed over the years, using a combination of traditional business-as-usual approaches and more sophisticated alternative assumptions, including a few backcasting estimates. In recent years, understanding the limitations of traditional water policies for solving unmet needs for water has encouraged new thinking about both methods for making projections and appropriate ways to meet anticipated needs for both water and water services. There is a growing interest in exploring more positive futures, developed using specific targets, objectives, or sustainability criteria and then exploring the ways of reaching those positive futures through specific policies or technologies.

When this approach is used, demands for water in the future are no longer assumed to rise inexorably. The earliest scenarios were largely disconnected from hydrologic reality, but newer efforts are more clearly tied to specific societal needs, wants, realistic hydrologic conditions, and concrete actions. When this approach is used on a regional scale it becomes more difficult to simply assume continued unsustainable use of groundwater or unmitigated water-quality contamination, because the "future" that results will fail to satisfy societal desires for a positive outcome.

New tools and approaches are becoming available to permit more sophisticated water planning and projections, and more detailed use of backcasting and other scenario concepts. Some of the analysis presented in earlier chapters for the food and agriculture, energy, and ecosystems sector represent new approaches for broadening the tools that can be applied to meet human needs while not destroying local and regional watersheds. When integrated with advances in our understanding about the potential for improving the productivity of industrial, commercial, and residential water use—and for expanding non-traditional sources of water supply, such as wastewater recycling and reuse, desalination, local rainwater harvesting, and conjunctive use of surface and groundwater—a "soft path for water" can not only be envisioned but perhaps more quickly reached. We are not predicting that such a future will come true; we are not assuming that our vision represents a global consensus. And indeed, it is unlikely to come about without fundamental changes in local, national, and international water policies and priorities. But such a future is both desirable and we believe achievable given existing technologies, strategies, and policies already being tried in various regions. As this chapter emphasizes, society will need to come to a general consensus of what we want to see when we reach 2050, and then define and implement measures that will take us to that vision. Otherwise, business-as-usual will not get us there and we, humanity, will be much the worst for that.

Our Water Future: Leadership and Individual Responsibility

Each chapter in this book has demonstrated how unsustainable it is to continue on our present path that requires more and more water from our limited resources to meet human needs. In some regions, demand already exceeds the sustainable volume available. Water demands exceeding supplies will become increasingly common. At the same time, our waste is polluting and contaminating the very resource on which we and other species of the ecosystem depend. Along with this, the inequitable distribution of the benefits of water resources leaves more than a billion people in ill health and absolute poverty. But it does not have to be like this. This book has shown the actions that can be taken across economic and social sectors to meet these multiple challenges. Examples are given of many actions already being taken around the world to deal with them. Leadership and individual responsibility can make our desirable future a reality. Together we can do it.

9.1 Water Management Challenges

Water is essential for our health, to grow our food, to provide our energy, and in general to support economic activity. Anyone who is aware of the water-related disasters in recent years, let alone the daily struggle of over a third of the world's population to obtain sufficient water of a quality that preserves their health, knows how the shortage, or excess, or pollution of water can affect us all. Parts of Africa, Australia, the Middle East, and North America are using far more water than can be replenished naturally. This is not sustainable. The impacts of global warming on our hydrologic cycle, water pollution, diminishing wetlands, the mismanagement of land as well as water resources, and the inequitable distribution of their benefits are among the main challenges currently facing humanity. How did we get to this situation?

Gulbenkian Think Tank on Water and the Future of Humanity, *Water and the Future of Humanity: Revisiting Water Security*, DOI 10.1007/978-3-319-01457-9_9,
© Calouste Gulbenkian Foundation 2014

9.2   Meeting Human Needs: Water-Related Challenges

The provision of these benefits comes from water resources, energy resources, and ecosystem services. Some progress has been made. For example, water managers working with others have made it possible to meet the needs of most of the seven billion people living today. About 90 % of the world population has access to improved sources of drinking water; an improvement of about 7 % during the 13 years since the United Nations' Millennium Development Goals (MDGs) were set. About 85 % are adequately nourished, although little progress has been made during this 13-year period.

Looking at the situation from another perspective, more than a billion people still lack access to safe drinking water, and about 2.5 million people lack adequate sanitation. Thirty-five million people die prematurely each year from water-related diseases. Three-quarters of the world's wastewater flows to the environment without treatment. A billion people have no electricity in their homes and a similar number lack clean fuel for heating and cooking. A billion live in absolute poverty. Meeting the needs of those without these benefits and of the estimated two billion more to be added to the population by 2050, poses many challenges (IWMI 2007; WWAP 2012; WWF 2011).

The following sections summarize the challenges described in earlier chapters for each major sector of water use.

9.2.1   Food Security

One of every seven people on the planet is chronically hungry. This means that they suffer from more disease, die younger, have reduced capacity to learn caused by diminished cognitive capacity (especially important in children), and are less productive if indeed they can find employment. In general, they are poor people located in places where the population is still growing and where the effective demand for food and their political bargaining power is weak. For people to lead active and healthy lives requires that food be available, that they have access to it, that the supply be stable, that it be safe to eat, and that they make healthy use of it. Those most urgently in need are small farmers who do not grow enough for their needs or earn enough to buy the food they need, and the urban poor without land who lack money to pay for food to properly feed themselves. Unfortunately, much of the food grown is lost through rot in storage or through other poor handling practices between the field and the market. More is wasted before it is eaten. Against this background, one in five people eat more than they require for healthy diets, are overweight, and subject to similar risks to those without access to enough food.

This is to a large extent a water challenge because food insecurity is mostly in regions where the availability of water and other resources needed for food production is constrained. In South Asia and the Middle East and North Africa (MENA) region rainfall is almost exhausted. Groundwater levels are critically low in parts of

China, India, Iran, Mexico, MENA, and the United States. More than 1.2 billion people live in river basins that are "closed," i.e., where all readily available water is already committed and allocated, and the trends of increasing shortages are serious. Looking to the future the challenge grows. At the World Food Summit in 2009, the recommendation was to increase food production (net of grains used for biofuels) by 70 % by 2050 to eliminate hunger and feed a population of nine billion. This implies a substantial increase, actually in excess of the rate of population increase, and will require significant efforts, investments, and additional inputs of water and energy. This is especially so if expansion of the area used for agriculture is to be minimized. The composition of agricultural production will also change. Current agricultural production is 62 % for humans, 35 % for animal feed (also producing food for humans), and 3 % for biofuels and other industrial uses. In North America only 40 % of agricultural land is dedicated to direct food production, while in Africa it is 80 %. With economic development, the percentage of production for animal feed tends to increase. Crop production for biofuel use is expected to more than triple by 2050.

At the same time there will be growing demand for energy from other sources and for rapidly expanding cities and industries. Climate change, economic instability and disparities, political instability, higher food prices, and rapid population growth, some occurring simultaneously, will add another dimension of uncertainty to the challenge.

9.2.2 Energy Security

Humans, their economies, and their societies critically depend on energy in the form of electricity and liquid and gaseous fuels. The principal uses are heating and cooling (34 %), transportation (28 %), and industry (27 %) (IIASA 2012). Domestic and public use accounts for the balance. Most of the current demand is being met. Yet, as noted earlier, a billion people have no electricity in their homes and a similar number lack clean fuel for heating and cooking. While the variety of energy uses makes it difficult to forecast future energy requirements as confidently as those for food, the estimates are that about 75 % more energy will be required by 2050 to serve those currently in need and meet future demand from population and economic growth.

The source of all energy on Earth is the Sun. Its heating of the Earth's atmosphere drives the hydrologic cycle, lifting water evaporated by the oceans and plant life, and transporting it so that it later falls on Earth on high altitudes from where its energy may be abstracted as hydropower as it flows to lower altitudes. Over millennia, ecosystems absorb the Sun's energy to grow and, later, after decaying, accumulate, being buried in the Earth's crust to become the source of peat, coal, natural gas, and methane. But turning these natural resources into electricity and useable solid and liquid fuel requires water. Growing biomass purposely for fuel and converting it to energy may consume 70–400 times as much water as producing energy from other sources. The production of energy is the second largest user of water; in the EU, 44 % of the water withdrawn is used for energy production.

Future water requirements to produce energy will depend on the technologies selected for energy production.

There is a special relationship between water and energy. The collection of water and its treatment, transmission, and distribution, followed by similar processes for wastewater, also requires energy. This presents both challenges and opportunities. When there is not sufficient water to meet all demands at acceptable costs, there may be a shortage of energy due, in part, to a shortage of water to produce it. Energy production can be planned to minimize water use. But if this increases the cost of energy, it will also increase the cost of water management. Using less energy for water management can save water used for energy production and at the same time reduce greenhouse gas emissions. The heart of the challenge is that energy producers in general do not control water resources management, and water management institutions have no control over energy production.

9.2.3   Urban and Industrial Demand

Today half of the world population lives in urban areas. UN Habitat predicts that virtually the entire demographic growth of the world over the next 30 years will be concentrated in urban areas. By 2050 urban dwellers will account for about 85 % of the population of the more developed world and some 65 % in the less developed regions—seven out of ten people will live in urban settlements. With this migration comes a complex trade-off between environmental risks in rural and urban settings. Migrants from rural areas often leave behind unsafe water supplies that put them at risk of waterborne infectious diseases, but they are exposed to new risks, such as urban air pollution, exclusion from health care, and poor housing conditions and their related communicable diseases.

Urban centers depend on water and energy inputs to function properly. The treatment and final disposal of liquid and solid wastes are still challenging the public health and public works agencies of the majority of urban centers, primarily in the developing world. Water supply, sanitation, wastewater treatment, storm water drainage, and solid waste management have been planned and delivered largely as isolated services. A range of authorities, each guided by distinct policies and pieces of legislation, continue to oversee these water subsectors at the city level.

Because the traditional urban water management model has failed to distinguish between different water qualities and identify uses for them, high-quality water has been diverted to indiscriminate urban water needs. Even basin-level management often neglects to acknowledge the cross-scale interdependencies in freshwater, wastewater, flood control, and storm water. Thus, there is a need for bringing forward an environmental approach that requires drastic measures of ecological rehabilitation, innovative institutional mechanisms, and a balance between autonomy and cooperation. Discussions of future sustainable urban water planning, development, and management now emphasize new strategies that are needed because water problems are becoming highly complex and interlinked with the urban supply of energy, food, employment, transportation, and job creation.

Integrated urban water management in the future will reframe a city's relationships to water and other resources, and reconceptualize the ways in which they can be overseen. Such global approaches include improving environmental monitoring and information by expanding the factual basis of comprehensive urban water management models. In addition, they imply a framework for negotiations that includes all of the stakeholders and stresses the importance of gradual but comprehensive institutional formats and clarity in local and national decision-making processes. Decentralized systems are sometimes proposed, as opposed to the centralized ones adopted by most cities in developed nations. But it is not an either/or option. Local systems are, and must be, part of larger physical and institutional contexts. The choice should be based on economic analysis, but also consider other aspects, such as institutional issues around responsibility for operation and maintenance and flexibility in system development.

Thanks to superior productivity, urban-based enterprises contribute large shares of gross domestic product (GDP). While in the past, industrial development has used as much water as might be available, today it is increasingly recognized through market signals that water has a value and that there is an opportunity cost associated with most uses of finite resources. As a result the trend is clear. Less is withdrawn and more wastewater is treated, so that both the water and some of the elements that accumulate in the production process can be reused. Industries in many countries are now consuming less water per unit output and reducing pollution loads in their waste. Water saving devices in households and elsewhere offer similar opportunities with lower net cost. For instance, domestic wastewater can be treated to separate valuable nutrients and reused together with the treated water for urban agriculture.

9.2.4   Resource Security

Water is the bloodstream of the biosphere as well as the source of our human livelihood; it links society and Nature. Because the amount of available freshwater in the biosphere is limited, and the demands for freshwater abstractions are increasing, there are increasing trade-offs and conflicts between off-stream water users and the in-stream uses of water to protect ecosystems. Fortunately, over recent years there has been increased recognition that water used to maintain the environment, or ecosystem integrity, also supports human needs by providing various services that benefit people. But the situation is still deteriorating. Because the water cycle is a biophysical and chemical process, inflowing water quality is as important as water quantity. The impact of freshwater scarcity on both human water security and ecosystems (see Fig. 4.1) calls for both global and site-specific remediation actions. The poor without food, energy, and other services are more likely to be dependent on food and fuel derived from their local environment and ecosystem. At the same time it is likely that a poorer person is involved in degrading the very ecosystem he or she is dependent upon.

We humans are not just spreading over the planet, but are changing the way fundamental geophysical and biophysical systems work. Whether we live in rural areas or cities, we manipulate vegetation, together with the soil and water to better meet our needs and wants. We are changing the landscape and adding contaminants. The impacts of such changes on the hydrologic cycle may be more profound than the impacts attributable to climate change. Part of the challenge is that taking actions to mitigate such changes takes time because society has to become aware of a problem before it can consciously respond.

Another challenge is that, as these dangers increase, important decision makers are not aware of, or ignore, the accumulated, if inadequate, knowledge about water flows and quality. Water observation networks, unfortunately, provide incomplete and incompatible data on water and wastewater quantity and quality for managing water resources and predicting future needs. Added to this is the stochastic character of water availability as a result of climate variability and the difficulties in predicting water flows and seasonal duration. To convert data to useful information in real time for various users is an enormous task. There is little sharing of useful hydrologic data, due largely to limited physical access to data, policy, and security issues, lack of agreed protocols for sharing, the huge data conversion task, and commercial considerations.

Furthermore, we still do not know enough about the interfaces between water and the soil, and the organisms found within the water and soil, and the abiotic and biotic processes that ensure important ecosystem services, such as the removal of water contaminants and the provision of clean drinking water. Nevertheless, we know enough to better manage our aquatic ecosystems. If we do not we are at risk of passing critical thresholds or tipping points beyond which a relatively small perturbation can qualitatively alter, perhaps irreversibly, the state or development of a system.

Finally, failure to value fully the multiple benefits that water and ecosystems provide through their different uses is a root cause of the public and political neglect of water and its mismanagement. Deriving the economic value of water in its different uses is controversial, has high data requirements, is complex, and requires technical and economic skills. While stakeholders represent some values, others have no voice.

9.2.5   Meeting Human Water Demands in 2050

Chapter 2 describes how human demand for water could evolve over the next 40 years under different conditions of demographic and economic growth, while taking account of climate change and environmental flow requirements. The results are presented by geographic region and economic sector in Figs. 2.7 and 2.8 under any condition, depending on the efficiency of water use and crop production and efficiencies in supply chains. Yet, as noted earlier, there are regions in the world where water resources are already stressed and where the ratio between people and resource endowment is worsening and compounded by higher risk. Where the demand exceeds supply, trade-offs will have to be made between human uses of water, between human use and environmental sustainability, and between what goods will be produced locally and what will be imported. But that will not be enough.

Exchanges with other regions in terms of trade, technology, and investments will be required. But desirable changes and exchanges do not come easily or by themselves. Many of the advances in water development and use have yielded significant benefits for millions of people, but not for all. For the future, the challenge is to cope with significant unresolved social challenges in a context where water is more uncertain and where further exploitation is much less of an option as compared to the recent past.

One important change in the future is a likely shift toward a much higher water demand for household and industrial uses in urban areas. Much of the water use in these sectors is considerably less consumptive as compared to the open landscapes of agriculture, where the resource is depleted through evaporation and transpiration. But storage capacities and conveyance systems are required. There are already many examples of huge transfers of water to urban and industrial uses. Even if a large fraction of the water used in urban centers returns to aquifers, lakes, or rivers, this will be in downstream areas affecting the quality. Further, many of the fast growing urban centers are located in coastal zones, which reduce options for reuse in other sectors at other locations.

Based on the assumption that economic growth together with demographic changes are key drivers behind increasing GDP, it is logical to expect that rapid improvements in GDP will be accompanied by increasing water demand. Certainly this occurred during the previous century (see Fig. 2.2, page 000). It is also happening with regard to recent rapid economic development in, e.g., China. At the same time, it is equally true that economic development is associated with innovations and more efficient resource use, captured by the term "more crop per drop." To what extent innovations and wide application of promising technologies and improved governance will reduce the rate of increase in demand in the future is not known. We can only guess. Figure 8.1 (see page 000) illuminates that projections made in the past about the trajectory of demand for water in the future have, so far, been exaggerations. This is important to keep in mind when interpreting the calculations presented in Figs. 2.6 and 2.7.

Yet, there is reason to assume that an inevitable increase in water demand will result in growing competition among water demand sectors in society, between urban and rural areas, and between sectors of the economy and the requirements for environmental flow. It is also well to recognize that improved efficiency in resource use does not necessarily lead to resource conservation, but may, rather, stimulate an expansion in aggregate resource exploitation through the rebound effect (Jevons' paradox). The challenges for policy are, therefore, to combine efforts that stimulate more efficient resource use with a governance package that ensures that potential resource savings from enhanced resource use efficiency will be used for increased allocations to cater for urgent needs that include the environment.

Efficiency must not only be interpreted in relation to production but also to what extent the goods and services produced reach their intended use. If a fraction of the goods and services produced are lost and wasted, it implies low efficiency in the supply chain. A high efficiency in production may thus be offset by a low efficiency in the supply chain. This is a significant challenge in the food sector, especially in connection with irrigated agriculture to which about 75 % of our freshwater is allocated and consumed. Reducing losses and waste in the food supply chain hinges on opportunities to save precious water for which there is a high opportunity cost. It may reduce the need for additional heavy investments in water infrastructure, increase the income of farmers through better access to markets, and reduce greenhouse gas emissions. In the end, losses and waste along the supply chain incur multiple costs and higher prices for consumers.

9.3   Climate Change and Governance

Meeting increasing demands for limited water resources and preventing pollution of this same resource are significant challenges in themselves. But climate change is affecting both the resource and the uses of the resource. A bigger challenge than climate change is that present governance systems are not designed to manage the complex system of competing water uses.

9.3.1   Climate Change

Climate variability and change have an important impact on water availability, safety, and ecosystems. Millions of people are affected every year by hydrological extremes, such as droughts and floods. Because of climate change, the number of and magnitude of extremes is likely to increase. Extreme events may span several years with devastating consequences, as in the Murray Darling Basin 2002–2009 (see Chap. 2) and the Sahel drought in West Africa in the 1980s. The changes will not be consistent across all regions. In some parts of the globe, water availability will increase while in other parts the water available for human use and ecosystems is likely to decrease. How climate change will affect water availability and extremes is still highly uncertain, making it difficult to integrate climate change information into water management.

The global climate has always been variable and climate change occurs at many different spatial and temporal scales. Temperature, rainfall, and other climate variables are continuously changing at global, regional, and local scales. The recent conclusion that humans are very likely affecting the global climate has initiated concern and research on how vulnerable we are to changes in climate systems and how we should adapt. It is important to distinguish between time scales. The synoptic time scale is where individual weather events may affect hydrological extremes, such as floods. The seasonal time scale affects water management through consistently higher or lower precipitation, which may result in floods, droughts, or seasonal water shortages. Seasonal weather patterns and forecasts are important for agricultural planning and management. At the decadal timescale, a projection of higher or lower than average precipitation could assist water resources planning measures, such as making adjustments to policy and infrastructure. Information on changes at the multi-decadal to century time scale can assist in designing and planning water infrastructure related to storage, water supply, and flood protection.

Because rainfall is often much more variable at annual timescales than temperature, it is harder to find trends in observations, and statistically significant trends in rainfall cannot be found for most places on the globe. There are some important exceptions. In Western Australia, e.g., rainfall has reduced by 10–20 % over the last 30 years, with a reduction in river discharge and dam inflows of more than 50 %. In the Pampas region of Argentina an opposite trend has been observed. Rainfall there has increased in some stations by more than 20 %. Because of the increased rainfall, land that was previously only suitable for rangeland, given its limited water availability, can now also be used for crop farming. Several stations in northern Eurasia also show a clear increasing trend in precipitation, resulting in higher river flows to the Arctic. Many river systems around the world depend on snowfall and glacial melt for significant parts of their runoff. Observations of glaciers and snow cover show reducing trends over the last 50 years over most parts of the globe.

Extreme events are, by definition, rare and in many regions there is large natural variability in the occurrence of extreme rainfall events, often making it difficult to find clear trends. On a global scale, it is clear that the frequency of extreme rainfall events has increased over recent decades and the number of intense hurricanes (Categories 4 and 5) also seems to have increased during this period. Examples of regions where more extreme events have been observed are southern Africa and northern Australia. The total number of floods, and the economic losses related to floods, has sharply increased during recent decades although it is still unclear what the role of climate change has been in the higher number of floods. Similarly, the intensity and duration of droughts have also increased since the 1970s. This is particularly the case in the tropics and subtropics. The increase in droughts is caused by a combination of reduced rainfall and higher temperatures. The Sahel region has suffered from more intense and longer droughts during the last 30 years.

Climate change will not only affect water resource availability and its variability but will also have an impact on water demand and use. With warmer temperatures and, therefore, greater evaporation, humans and animals will require more water. More water will also be required for domestic, energy, industrial, and agricultural uses. This will arise from the greater cooling requirements, greater evaporation, and higher water temperatures. Any change in agricultural water demand due to climate change can have a large impact on water resources. Crop water demands will change with higher temperatures, but growing seasons will also get longer so that higher yielding crop varieties can be used or more crops can be grown on the same land in 1 year (multi-cropping). The areas where crops are grown and the types of crops grown will change and, therefore, change the water demands. This will also be true of natural vegetation, ecosystems, and species, which will also change as the climate changes. In many cases higher latitudes will become more suitable for agriculture and dry areas become less suitable.

It is now recognized that to take account of the uncertainties and risks associated with climate change and other unknown factors better water management will depend on an analysis of scenarios of possible futures. Stakeholder groups from many different sectors and disciplines, including those that need to make planning decisions in their sectors, should be involved in developing a range of scenarios which are both deemed plausible by the many different sectors and of interest and useful to the relevant users. Scientific experts in the different disciplines ensure that the scenarios developed are credible scientifically, help to quantify what the scenarios would mean, and communicate that information back to stakeholder groups for further refinement, improved communication and understanding, and use in decision making.

9.3.2 Appropriate and Effective Water Governance

"Water governance refers to the range of political, social, economic and administrative systems that are in place to regulate development and management of water resources and provisions of water services at different levels of society" (UNDP 2011). Many problems are not primarily associated with the resource base, but can be attributed to governance failures and poor understanding of how natural and economic systems function. Human institution interactions affect water use; the human dimension, thus, has a central role and strong emphasis needs to be given to governance issues.

Until recently, to meet human needs, water managers focused on the management of water alone. Managers of other resources neglected water. Users of water focused on their own needs, with little concern for other users. Few paid attention to the ecosystem as a resource or a user of water because it had no representative to speak for it. Since near the turn of the century, sustainability of the resource has become a part of integrated water resource management and this includes ecosystems. Management of resources, including water, has not yet adequately addressed human welfare. For example, world food production is more than sufficient to feed the global population, yet nearly one-third of the population suffers from illness caused by either starvation or overeating. Little attention is paid to what may be called the vertical integration, from production to access and end use of the product. Our institutions are failing to ensure the appropriate and equitable distribution of resource development. A combination of horizontal and vertical integration would do this, as illustrated earlier in Fig. 2.8.

More recently, awareness has been growing that the major decisions taken that impact water resources and water use are taken outside the "water box." The analyses in this book have demonstrated the linkages among all uses and all natural resources, and that balancing them is a social and political challenge requiring the appropriate institutions. In North America, severe threats to the world's largest freshwater ecosystem grassroots made public participation essential for achieving the binationally set objectives to restore and protect the ecosystem. Strategies to improve the "ecosystem quality" of the Great Lakes were developed by the International Joint Commission through a public participation process involving industrialists, small business people, farmers, labor representatives, educators, environmentalists, and representatives of women's groups, sportsmen's and fishermen's associations, wildlife federations, extension agents, and elected and appointed officials (Becker 1993).

Most institutions of governance are now recognizing the benefits of involving stakeholders in their planning and decision-making processes. Yet this is not easy, nor is it efficient in the short run. It takes time, patience, and skill. As a consequence, there often continues to be inadequate equitable and effective participation of stakeholders in decision making, and decisions are taken without the benefit of understanding the consequences for other resources and other users. Improved data collection, interpretation, and coordination can result from developing personal, working relationships between utility or agency managers. Proprietary issues and other barriers to trust can arise less frequently when stakeholders are personally familiar with each other. The special relationship described in Chap. 7 between energy and water is a clear demonstration of the need for coordinated institutions. The lack of connection between water and energy policy is driven, in large part, by the failure of water and energy practitioners to engage with, and fully understand, one another and the issues each must deal with.

The many conferences on the energy, water, and food nexus during preparations for the RIO+20 Conference, held in Rio de Janeiro in 2012, were recognition of this need. The challenge is to create such new institutions at, and between, all levels of governance, following the principle of subsidiarity. Science, including the social sciences, is challenged to design systems that permit integrated analyses to support decision making and present the results in a manner to facilitate stakeholder participation in making trade-offs.

9.4   Responding to Water-Related Challenges

Water plays a pivotal role in the functioning of the global Earth System. It is also an essential ingredient for the life of humanity, for human activities, and for supporting ecosystems. Global, regional, and local water issues will tend to become increasingly complex and access to water will become progressively unequal in various regions of our planet. Globalization, the action of international institutions and inter-governmental cooperation organizations, the increased value of human and natural capital, and the prodigious development of science and technology, particularly of information and communication technologies, are external factors that will play a major role in how we manage water in the future. If we pursue the prevailing business-as-usual approach, more serious problems will tend to arise. Water management may become a key issue for society. Among the aspects that require careful analysis is the validation of current water management paradigms. When justified, water managers must question conventional wisdom and practice, current management methods, the availability of reliable data, and the adequacy of the technical–scientific knowledge.

Water managers will have to handle more complex future challenges and respond to a much wider range of issues than they have been used to in the recent past. The complexity of water issues stems, in a large measure, from increasingly important interactions between water, energy, food, and ecosystems. There is growing awareness, reflected in the media, that the path we are following is unsustainable. But there is inadequate awareness that water links all socioeconomic sectors and goals and that cooperation among them is essential. Water managers can alert society to the complexity and difficulties of the water-related problems that lie ahead and, thereby, generate the political will needed to successfully cope with them.

When we consider the present water problems, we often tend to think first of the basic water needs related to water supply and sanitation. The UN MDGs considered this as a primary issue and aimed to halve, from 2000 to 2015, the proportion of the world population without access to improved sources of drinking water and basic sanitation. The world is on track to meet the drinking water target, though much remains to be done in some regions. Unfortunately, sanitation is still lagging behind. Basic water supply and sanitation can be provided at low-cost as the World Health Organization (WHO) and others have amply demonstrated. External assistance is readily available for this purpose for countries with the will and capacity to do some something about it. Civil war and other conflicts mean that some countries have even higher short-term priorities.

However, the reality is that major investments are needed to modernize and efficiently manage the existing water and sanitation services and to build and operate new water infrastructures. John Briscoe, for 10 years Senior Water Advisor at the World Bank and in his last assignment Country Director for Brazil, said in an interview (Briscoe 2011):

> Every country which has successfully lifted its people out of poverty has done so primarily by building its basic productive capacity. Central to this process has been giving priority to improving the productivity of agriculture, and creating the energy, transport and water infrastructure for rural and urban economic growth and employment generation. No presently-rich country has developed without such investments, which have been the springboard for private sector growth, for job creation, for agricultural productivity. To take but one indicator every presently-rich country has developed more than 70 % of its economically-viable hydro-electric potential. Africa has developed 3 % of its potential. Not only is this the path that has been followed by all presently-rich countries, but it is the path followed by the countries who have, in recent decades, pulled their people out of poverty—like China, India and Brazil. Of course infrastructure is not a sufficient condition for poverty reduction, but it most certainly is a necessary condition!

The situation in the future may become much more difficult considering the issues facing us. Citizens of the emerging countries, such as China and India, aspire to accede to standards and life styles comparable to those of the developed countries. We can well expect a high increase not only of the direct water consumption per capita but also of the water consumed in the production of different consumption products, particularly food products. These countries' populations correspond to more than one-third of the world's population. Size matters. Water, although renewable, is definitely a limited resource and will become limiting in many countries. It has no substitute.

Trade-offs between satisfying increased human demands for water and reducing the adverse impacts to ecosystems services are unavoidable. They will require the balancing of social benefits to be satisfied and environmental services to be preserved. However, opportunities to provide people with more goods and services can be enhanced without a corresponding increase in production. Policies aiming at these kinds of improvements would make it possible to meet the needs and wants in society in a more resource-efficient manner. For example, as discussed in Chaps. 2 and 6, 30–50 % of food produced in the field is lost in post-harvest operations, in processing, and in waste at the end of the food production chain. Perhaps less, but still important, reductions are possible in the use of industrial and household water.

Measures taken to reduce poverty and promote economic growth and development need to work together in a synergistic manner. High-level political leadership is necessary for that to happen. World Water Development Reports 3 and 4 provide examples of how water managers, governments at several levels, and the private sector are already addressing the challenges (WWAP 2009, Chaps. 14 and 15; WWAP, 2012, Chaps. 13 and 14). There are examples of more integrated and better-coordinated approaches to water, land use, and ecosystems, addressing the role of water in the context of socioeconomic development and environmental sustainability. Quantified analyses of possible pathways to an agreed desirable future can be, and are being, developed and will make it possible to identify actions that are positive and robust under any future scenario. In the Great Lakes case described above, new participatory software allowed political and technical stakeholders to jointly create simulation models that have become trusted monitoring tools of five sovereign entities (Delli Priscoli 2004). This reflects the need to monitor, review, and modify or adapt actions being taken to stay on track toward achieving a desired future.

9.5   Leadership, Commitment, and Responsibility

We believe that in a desirable future for humanity, all people would at all times have secure, affordable, and adequate access to:
- Safe water to meet personal and household needs
- Safe nutritious food for a healthy and active life
- Clean energy for heating, cooling, and transportation needs
- Healthy housing and a surrounding environment

This would be supported by a governance system in which everybody contributes in an equitable manner to making decisions that affect lives and livelihoods. The analysis of major development needs and their dependencies on water and other resources, as we have shown in the previous chapters of this book, reveals technological, economic, political, and social options that, if adopted, would help us to meet these needs. One option is to improve water use efficiency. There is a large space for much improvement, particularly in water use for agriculture. Encouraging the development of creative abilities is essential and urgent, not only in approaches to technology and economics (where there is some interest) but also in reforming political and governance systems. Dealing efficiently with the problems of water in all their complexity and the implementation of economically efficient, socially equitable, and environmentally sustainable water resources management can only be achieved through good water resources governance.

A large number of people on the planet have achieved the MDGs and some have surpassed them, albeit in some cases in an unsustainable manner. Consider the progress that can be made if there is agreement on goals, as with the MDGs. Consider specifically the MDG to halve the number of people without access to safe drinking water by 2015. Because defining "safe drinking water" and monitoring it would be difficult, a target was built on "access to an improved source of drinking water." In the 20 years from 1990, over two billion people were provided with such improved access. If this rate of progress is maintained, all will be served well before 2050. Thus, an agreed, well-defined goal can be achieved with the will, the effort, the technical capacity, and the funds.

As demonstrated in this book, new tools and approaches are becoming available to permit more sophisticated quantitative water planning and projections by applying an integrated systems approach. Scientific knowledge and technological development are expanding exponentially. Today they are limited only by our imagination. New communications technology makes it possible for us to not only share knowledge (collective intelligence) but also to determine our shared values and objectives and a shared strategy to act on them. Increasingly women and young people are making their voices heard and participating in decision-making processes. We can work together to build the sustainable world we agree to seek. While challenges are great, together we can do it. It will require leadership, individual responsibility, and commitment to action.

To conclude, consider this quote from Jacques Cousteau, as cited in an interview of Ted Turner (President of the UN Foundation) by *Variety* magazine (August 3, 2012): "Even if we knew for sure we were going to lose, which we don't—what else would men of good conscience do, but do their best to the very end, keep fighting to the very end."

Many who have worked in the water field over the years have contributed significantly to human development. All who we know are committed to working even more closely with other partners to continue to make the planet a better place for all—beginning not tomorrow, but now.

References

ABARE (Australian Bureau of Agricultural and Resource Economics). (2010, December). *Australian wheat supply and exports monthly*. Canberra, ACT, Australia: ABARE.

Abbott, A., Hurt, C., & Tyner, W. (2011). *What's driving food prices in 2011? Issue report*. Oak Brook, IL: Farm Foundation.

Adam, J., & Lettenmaier, D. (2008). Application of new precipitation and reconstructed streamflow products to streamflow trend attribution in Northern Eurasia. *Journal of Climate, 21*(8), 1807–1828.

ADB (Asian Development Bank). (2010). *Strengthening the resilience of the water sector in Khulna to climate change*. Manila, Philippines: ADB (Final Report ADB TA–7197).

Ait Kadi, M. (1997, August 16). High water stress–low coping capacity — Morocco's example. In *Proceedings of the Mardel Plata 20 year anniversary seminar*. Stockholm, Sweden: Stockholm International Water Institute.

Ait Kadi, M. (2000, November 20–22). Les Politiques de l'Eau et la Sécurité Alimentaire au Maroc à l'Aube du 21ème Siècle — Exposé Introductif. Publications de l'Académie du Royaume du Maroc. *Session d'Automne*, 33–75. Rabat, Morocco (In French).

Ait Kadi, M. (2009a). *Impacts du changement climatique sur la sécurité alimentaire. Acts of the international meeting on adapting to climate change in Morocco*. Rabat, Morocco: Royal Institute for Strategic Studies, pp. *95–108* (In French).

Ait Kadi, M. (2009b.) La crise alimentaire mondiale 2007–2008. *Bulletin de l'Académie Hassan II des Sciences et Techniques* N°5. Rabat, Morocco (In French).

Alcamo, J., Döll, P., Kaspar, F., & Siebert, S. (1997). *Global change and global scenarios of water use and availability: An application of WaterGAP 1.0*. Kassel, Germany: Wissenschaftliches Zentrum Für Umweltsystemforschung, Universität Gesamthochschule.

Alcamo, J., Flörke, M., & Märker, M. (2007). Future long-term changes in global water resources driven by socioeconomic and climatic changes. *Hydrological Sciences Journal, 52*, 247–275.

Alexandratos, N., & Bruinsma, J. (2012). *World agriculture towards 2030/2050: The 2012 revision*. Rome: FAO (ESA Working paper No. 12–03).

ANA (Agencia Nacional de Aguas). (2010). *Atlas de Abastecimento de Água do Brasil*. Brasilia, Brazil: ANA (In Portuguese).

Angel, S., Daniel, J., Civco, L., & Blei, A. M. (2011). *Making room for a planet of cities*. Cambridge, MA: Lincoln Institute of Land Policy.

Arent, D. J. (2010). The role of renewable energy technologies in limiting climate change. *The Bridge, 40*, 31–39.

ASCE (American Society of Civil Engineers. (2009). Report card 2009 grades. In *Report card for America's infrastructure*. Reston, VA: American Society of Civil Engineers. Retrieved from http://www.asce.org/reportcard/2009/grades.cfm

ASCE (American Society of Civil Engineers). (1998). *Sustainability criteria for water resource systems*. Reston, VA: ASCE Press.

ASCE (American Society of Civil Engineers). (2011). *Toward a sustainable water future: Visions for 2050*. Reston, VA: ASCE Press.

Asseng, S., Travasso, M. I., Ludwig, F., & Magrin, G. O. (2013). Has climate change opened new opportunities for wheat cropping in Argentina? *Climatic Change, 117*(1–2), 181–196. doi:10.1007/s10584-012-0553-y.

Atlantic Council. (2011). *Energy for water and water for energy: A report on the Atlantic Council's workshop on how the nexus impacts electric power production in the United States.* Washington, DC: Atlantic Council.

Averyt, K., Fisher, J., Huber-Lee, A., Lewis, A., Macknick, J., Madden, N., et al. (2011). *Freshwater use by US power plants: Electricity's thirst for a precious resource.* Cambridge, MA: Union of Concerned Scientists (Report of the Energy and Water in a Warming World Initiative).

AWWA (American Water Works Association). (2011). High energy costs comprise half of some city budgets. *Streamlines, 3*(10).

Bahri, A. (2012). *Integrated urban water management.* Stockholm, Sweden: Elanders (Global Water Partnership Technical Committee Background Paper No. 16).

Barnett, T. P., Adam, J., & Lettenmaier, D. P. (2005). Potential impacts of a warming climate on water availability in snow-dominated regions. *Nature, 438*, 303–309.

Bates, B. C., Kundzewicz, Z. W., Wu, S., & Palutikof, J. P. (2008). *Climate change and water.* Geneva, Switzerland: Intergovernmental Panel on Climate Change (IPCC Technical Paper VI).

Becker, M. L. (1993). The International Joint Commission and public participation: Past experiences, present challenges, future tasks. *Natural Resources Journal, 33*, 236–274.

Beddington, J., Asaduzzaman, M., Clark, M., Fernandez, A., Guillou, M., Jahn, M., et al. (2012). *Achieving food security in the face of climate change: Final report from the Commission on Sustainable Agriculture and Climate Change.* Copenhagen, Denmark: CGIAR Research Program on Climate Change, Agriculture and Food Security (CCAFS).

Ben Mabrouk, S. (2008). Les enjeux des ressources naturelles dans la péréquation: Cas des nouvelles redevances hydrauliques au Québec. (In French) Retrieved from http://archimede.bibl.ulaval.ca/archimede/fichiers/25746/ch04.html

Bergkamp, G., & Sadoff, C. (2008). Water in a sustainable economy. In W. Institute (Ed.), *State of the world: Innovations for a sustainable economy* (pp. 107–122). Washington, DC: Worldwatch Institute.

Board, L. W. S. (2006). *Reducing nutrient loading to Lake Winnipeg and its watershed: Our collective responsibility and commitment to action.* Winnipeg, MB, Canada: Government of Manitoba, Ministry of Water Stewardship.

Boyd, J., & Banzhal, S. (2006, January). *What are ecosystems services?* Washington, DC: Resources for the Future (Discussion Paper 06-02).

Braga, B., Filho, J. G. C. G., von Borstel Sugai, M. R., Vaz da Costa, S., & Rodrigues, V. (2012). Impacts of Sobradinho Dam, Brazil. In *Impacts of large dams: A global assessment* (pp. 153–170). Berlin, Germany: Springer.

Braga, B. P. F., Flecha, R., Thomas, P., Cardoso, W., & Coelho, A. C. (2009). Integrated water resources management in a federative country: The case of Brazil. *Water Resources Development, 25*, 611–628.

Briscoe, J. (2011). Invited opinion interview: Two decades at the center of world water policy: Interview with John Briscoe by the Editor-in-Chief. *Water Policy, 13*, 147–160.

Brocklesby, M. A., & Hinshelwood, E. (2001). *Poverty and the environment: What the poor say—An assessment of poverty-environment linkages in participatory poverty assessments.* CDS.: University of Wales, Swansea, UK.

Brooks, D. B., & Brandes, O. M. (2005). *The soft path for water in a nutshell.* Victoria, BC, Canada: Friends of the Earth Canada/POLIS Project on Ecological Governance, University of Victoria.

Brooks, D. B., Brandes, O. M., & Gurman, S. (Eds.). (2009). *Making the most of the water we have: The soft path approach to water management.* London, UK: Earthscan.

Brown, K., Daw, T., Rosendo, S., Bunce, M., & Cherrett. N. (2008). Ecosystem services for poverty alleviation: marine & coastal situational analysis: Synthesis report. Ecosystem services for poverty alleviation programme. University of East Anglia, Norwich, UK: ODG. Retrieved from http://www.nerc.ac.uk/research/programmes/espa/documents/Marine%20and%20Coastal%20-%20Synthesis%20Report.pdf

Bruun, C. (2010). Imperial power, legislation and water management in the Roman Empire. *Insights* 3 (10). Durham University, Durham, UK: Institute of Advanced Studies.

CALFED. (2006). Water use efficiency comprehensive evaluation. In *CALFED Bay—Delta Program water use efficiency element*. Sacramento, CA: CALFED Bay Delta Program.

California Department of Water Resources. (2009). *California water plan update*. Bulletin 160–09. Sacramento, CA: California Department of Water Resources.

California Energy Commission. (2005). *California's energy–water relationship*. CEC-700-2005-001-SF. Sacramento, CA: California Energy Commission.

Carlsson, G., Cropper, A., El-Ashry, M., Honglie, S., Hvidt, N., Johnson, I., et al. (2009). *Closing the gaps, report of the commission on climate change and development*. Stockholm, Sweden: Sweden Ministry of Development and Cooperation.

Carpenter, S. R., & Bennett, E. M. (2011). Reconsideration of the planetary boundary for phosphorus. *Environmental Research Letters, 6*(1).

Carter, N. T. (2010). *Energy's water demand: Trends, vulnerabilities and management*. 7–5700. Washington, DC: Congressional Research Service.

CAWMA (Comprehensive Assessment of Water Management in Agriculture). (2007). *Water for food, water for life: A comprehensive assessment of water management in agriculture*. London, UK/Colombo, Sri Lanka: Earthscan/IWMI.

Chartres, C., & Varma, S. (2010). *Out of water: From abundance to scarcity and how to solve the world's water problems*. Upper Saddle River, NJ: Pearson Education.

Choudhury, G. A., van Scheltinga, C. T., van den Bergh, D., Chowdhury, F., de Heer, J., Hossain, M., et al. (2012). *Preparations for the Bangladesh delta plan*. Wageningen, Netherlands: Alterra Wageningen.

Cicek, N., Lambert, S., Venema, H. D., Snelgrove, K. R., Bibeau, E. L., & Grosshans, R. (2006). Nutrient removal and bio-energy production from Netley-Libau Marsh at Lake Winnipeg through annual biomass harvesting. *Biomass and Bioenergy, 30*(6), 529–536.

Cohen, J. E. (2010). *Beyond population: Everyone counts in development*. Retrieved from http://www.cgdev.org/files/1424318_file_Cohen_BeyondPopulation_FINAL.pdf

Collier, U. (2006). Meeting Africa's energy needs—The costs and benefits of hydro power. Worldwide Fund for Nature, Oxfam and Wateraid joint report.

Collier, P. (2007). *The bottom billion: Why the poorest countries are failing and what can be done about it*. New York: Oxford University Press.

Cooley, H., Fulton, J., & Gleick, P. H. (2011). *Water for energy: future water needs for electricity in the intermountain west*. Oakland, CA: Pacific Institute.

Cosgrove, W. J. (2008). Public participation to promote water ethics and transparency. In M. Ramon Llamas et al. (Eds.), *Water ethics*. Leiden, Netherlands: CRC Press.

Cosgrove, W. J., & Rijsberman, F. R. (2000). *World water vision: Making water everybody's business*. London, UK: Earthscan.

Cosgrove, W. J., & Tropp, H. (2013). *Water for development: Investing in health and economic well-being globally in ensuring a sustainable future: Making progress on environment and equity*. Heymann J. and Barrera M. (eds.), New York: Oxford University Press.

Council, W. E. (2010). *Water for energy. Executive summary*. London, UK: World Energy Council.

Crutzen, P. J. (2002). Geology of mankind. *Nature, 415*, 23.

Crutzen, P. J., & Stoermer, E. F. (2000). The 'Anthropocene'. *Global Change Newsletter, 41*, 17–18.

CSIRO (Commonwealth Scientific and Industrial Research Organisation). (2008). *Water availability in the Murray–Darling Basin Report*. Clayton South, VIC, Australia: CSIRO.

Daigger, G. T. (2007, October 24–25). Creation of sustainable water resources by water reclamation and reuse. In *Proceedings of the 3rd International conference on sustainable water environment: Integrated water resources management—New steps* (pp. 79–88). Sapporo, Japan.

Daigger, G. T. (2008). New approaches and technologies for wastewater management. *The Bridge, 38*(3), 38–45.

Daigger, G. T. (2012). A vision for urban water and wastewater management in 2050. In W. M. Grayman, D. P. Loucks, & L. Saito (Eds.), *Toward a sustainable water future: Visions for 2050* (pp. 113–121). Washington, DC: American Society of Civil Engineers.

Daigger, G. T., & Crawford, G. V. (2007). Enhanced water system security and sustainability by incorporating centralized and decentralized water reclamation and reuse into urban water management systems. *Journal of Environmental Engineering Management, 17*(1), 1–10.

Daily, G. C. (Ed.). (1997). *Nature's services: Societal dependence on natural ecosystems.* Washington, DC: Island Press.

Daniel, S., & Mittal, A. (2009). *The great land grab; rush for world's farmland threatens food security for the poor.* Oakland, CA: The Oakland Institute.

Davies-Colley, R. J., Smith, D. G., Ward, R. C., Bryers, G. G., McBride, G. B., Quinn, J. M., et al. (2011). Twenty years of New Zealand's national river water quality network: Benefits of careful design and consistent operation. *Journal of the American Water Resources Association, 47*(4), 750–771.

de Wit, M., & Stankiewicz, J. (2006). Changes in surface water supply across Africa with predicted climate change. *Science, 311*(5769), 1917–1921.

DeCarolis, J., Adham, S., Pearce, W. R., Hirani, Z., Lacy, S., & Stephenson, R. (2007). Cost trends of MBR systems for municipal wastewater treatment. In Water Environment Federation (Ed.), *WEFTEC 07* (pp. 3407–3418). San Diego, CA: Water Environment Federation.

Delli Priscoli, J. (2004). What is public participation in water resources management and why is it important? *Water International, 29*(2), 221–227.

Deltacommissie (Delta Commission). (2008). Working together with water: A living land builds for its future. Amsterdam, Netherlands: Deltacommissie. Retrieved from http://bit.ly/bKtIHw

Dixit, A. (2009). Governance, institutions and economic activity. *The American Economic Review, 99*(1), 5–24.

Doorn, N., & Dicke, W. (2012). SPRAAKWATER: Values of water. *Water Governance, 2*(2), 53.

Dyson, M., Bergkamp, G., & Scanlon, J. (2004). *Flow: The essentials of environmental flow.* Gland, Switzerland: IUCN.

Edvard, C. (2011). *Technology innovation is everybody's business.* Electrical engineering portal. Retrieved February 18, 2012 from http://electrical-engineering-portal.com/technology-innovation-is-everybodys-business

Ehrhardt-Martinez, K., & Laitner, J. A. (Eds.). (2010). *People-centered initiatives for increased energy savings.* Boulder, CO: Renewable and Sustainable Energy Institute, University of Colorado/American Council for an Energy-Efficient Economy.

Europe's World. (2012, Summer). *Special section: Water; water and energy security are two sides of the same coin.* Retrieved from http://www.siwi.org/documents/Resources/news_articles/EW21water.pdf

Falkenmark, M., & Rockström, J. (2004). *Balancing water for humans and nature: The new approach in ecohydrology.* London, UK: Earthscan.

Falkenmark, M., & Rockström, J. (2006). The new blue and green water paradigm: Breaking new ground for water resources planning and management. *Journal of Water Resources Planning and Management, 5*(6), 129–132.

FAO (Food and Agricultural Organization of the United Nations). *AQUASTAT online database.*

FAO (Food and Agriculture Organization of the United Nations). (1996, November 13–17). *Declaration on world food security.* World Food Summit. Rome: FAO

FAO (Food and Agriculture Organization of the United Nations). (2001). Food insecurity: when people live with hunger and fear starvation. In *The state of food insecurity in the world.* Rome, Italy: FAO.

FAO (Food and Agriculture Organization of the United Nations). (2002). *Crops and drops: Making the best use of water for agriculture.* Rome, Italy: FAO.

FAO (Food and Agriculture Organization of the United Nations). (2003). Food security: Concepts and measurement. In *Trade reforms and food security.* Retrieved from http://www.fao.org/docrep/005/y4671e/y4671e06.htm

FAO (Food and Agriculture Organization of the United Nations). (2005). *Reducing fisherfolk's vulnerability leads to responsible fisheries: Policies to support livelihoods and resource management. New directions in fisheries policy briefs.* Rome, Italy: FAO.

FAO (Food and Agriculture Organization of the United Nations). (2007). *The state of world fisheries and aquaculture—2006.* Rome, Italy: FAO.

FAO (Food and Agriculture Organization of the United Nations). (2008). The state of food and agriculture. In *Biofuels: Prospects, risks and opportunities*. Rome, Italy: FAO.

FAO (Food and Agriculture Organization of the United Nations). (2009). *Crops prospects and food situation*. No. 2. Retrieved from http://www.fao.org/docrep/011/ai481e/ai481e04.htm

FAO (Food and Agriculture Organization of the United Nations. (2010). The state of food insecurity – Addressing food insecurity in protracted crises – Rome.

FAS, USDA (Foreign Agricultural Service United States Department of Agriculture). (2011). *Production, supply and demand online*. Washington, DC: USDA.

Fay, M., & Toman, M. (2010). *Infrastructure and sustainable development: Post-crisis growth and development* (pp. 329–382). Washington, DC: International Bank for Reconstruction and Development/The World Bank.

Feeley, T. J., III, Skone, T. J., Stiegel, G. J., Jr., McNemar, A., Nemeth, M., Schimmoller, B., et al. (2008). Water: A critical resource in the thermoelectric power industry. *Energy, 33*, 1–11.

Fencl, A., Clark, V., Mehta, V., Purkey, D., Davis, M., & Yates, D. (2012). Water for electricity: Resource scarcity, climate change and business in a finite world. In *Project Report 2012*. Stockholm, Sweden: Stockholm Environment Institute.

Fischer, G., Shah, M., & van Velthuizen, H. (2002). *Climate change and agricultural vulnerability*. Laxenburg, Austria: IIASA.

Fischer, G., Tubiello, F., van Velthuizen, H., & Wiberg, D. A. (2007). Climate change impacts on irrigation water requirements: Effects of mitigation 1990–2080. *Technological Forecasting and Social Change, 74*, 1083–1107.

Foley, J., Ramankutty, N., Brauman, K. A., Cassidy, E. S., Gerber, J. S., Johnston, M., et al. (2011). Solutions for a cultivated planet. *Nature, 478*, 337–342.

Foresight. (2011). *The future of food and farming—Challenges and choices for global sustainability. Executive summary*. London: GO-Science.

Gallopin, G. C. (2011). Five stylized world water scenarios. In *Global water futures 2050*. United Nations World Water Assessment Programme. Paris, France: UNESCO.

Gallopin, G. C., & Rijsberman, F. (2000). Three global water scenarios. *International Journal of Water, 1*, 16–40.

GAO (General Accounting Office). (2003). *States' views of how federal agencies could help them meet the challenges of expected shortages*. Washington, DC: GAO. Retrieved from http://gao.gov/products/GAO–03–514

GAO (General Accounting Office). (2009). *Improvements to federal water use data would increase understanding of trends in power plant water use*. Washington, DC: GAO.

Garneau, J.-Y. (2012). *Pour que la terre soit notre amie. L'aventure intérieure*. Montreal, QC, Canada: Novalis (In French).

Gerbens-Leenes, W., Hoekstra, A., & van der Meer, T. (2008). The water footprint of energy consumption: An assessment of water requirements of primary energy carriers. *ISESCO Science and Technology Vision, 4*, 38–42.

Gini, C. (1912). *Variabilità e mutabilità (Variability and mutability)*. Bologna, Italy: C. Cuppini (In Italian).

Giordano, M. (2009). Global groundwater: Issues and solutions. *Annual Review of Environment and Resources, 34*, 7.1–7.26.

GIZ (Deutsche Gesellschaft für Internationale Zusammenarbeit GmbH). (2001, December 4). *International conference on freshwater*. Bonn, Germany: GIZ.

Glassman, D., Wucker, M., Isaacman, T., & Champilou, C. (2011). *World policy papers: The water–energy nexus: Adding water to the energy agenda*. New York: WPI.

Gleick, P. H. (1996). Basic water requirements for human activities: Meeting basic needs. *Water International, 21*(2), 83–92.

Gleick, P. H. (1997). *Water 2050: Moving toward a sustainable vision for the earth's fresh water. Working Paper of the Pacific Institute for Studies in Development, Environment, and Security, Oakland, California*. Prepared for the Comprehensive Freshwater Assessment for the United Nations General Assembly and the Stockholm Environment Institute, Stockholm, Sweden.

Gleick, P. H. (2000a). A picture of the future: A review of global water resources projections. In T. World's (Ed.), *Water 2000–2001: The biennial report on freshwater resources* (pp. 39–61). Washington, DC: Island Press.

Gleick, P. H. (2000b). Water bag technology. In P. H. Gleick (Ed.), *The world's water 1998–1995: The biennial report on freshwater resources* (pp. 200–205). Washington, DC: Island Press.

Gleick, P. H. (2002). Soft water paths. *Nature, 418*, 373.

Gleick, P. H. (2003). Water use. *Annual Review of Environment and Resources, 28*, 275–314.

Gleick, P. H. (2009a). China and water. *The world's water 2008–2009: The biennial report on freshwater resources*, 79–100. Washington, DC: Island Press.

Gleick, P. H. (2009b). Getting it right: Misconceptions about the soft path. In D. B. Brooks, O. M. Brandes, & S. Gurman (Eds.), *Making the most of the water we have: The soft path approach to water management* (pp. 49–60). London, UK: Earthscan.

Gleick, P. H., Christian-Smith, J., & Cooley, H. (2011). Water-use efficiency and productivity: Rethinking the basin approach. *Water International, 36*, 784–798.

Gleick, P. H., Haasz, D., Henges-Jeck, C., Srinivasan, V., Wolff, G., Cushing, K. K., et al. (2003). *Waste not, want not: The potential for urban water conservation in California*. Oakland, CA: Pacific Institute for Studies in Development, Environment, and Security.

Gleick, P. H., Loh, P., Gomez, S. V., & Morrison, J. (1995). *California water 2020: A sustainable vision*. Oakland, CA: Pacific Institute.

Gleick, P. H., & Palaniappan, M. (2010). Peak water: Conceptual and practical limits to freshwater withdrawal and use. *Proceedings of the National Academy of Sciences of the United States of America, 107*(25), 11155–11162.

Gleick, P. H., Wolf, G., Chaleki, E. L., & Reyes, R. (2002). *The new economy of water: The risks and benefits of globalization and privatization of fresh water*. Oakland, CA: Pacific Institute for Studies in Development, Environment and Security.

Goldstein, N. C., Newmark, R. L., Whitehead, C. D., Burton, E., McMahon, J. E., Ghatikar, G., et al. (2008). The energy–water nexus and information exchange: Challenges and opportunities. *International Journal of Water, 4*(1/2), 5–24.

Goolsby, D. A., Battaglin, W. A., & Hooper, R. P. (1997). Sources and transport of nitrogen in the Mississippi River Basin. Retrieved from http://co.water.usgs.gov/midconherb/html/st.louis.hypoxia.html

Grey, D., & Sadoff, C. (2007). Sink or swim? Water security for growth and development. *Water Policy, 9*, 545–571.

Griffiths-Sattenspiel, B., & Wilson, W. (2009). *The carbon footprint of water*. Portland: River Network.

Gudmundsson, L., Tallaksen, L., & Stahl, K. (2011). Projected changes in future runoff variability—A multi-model analysis using the A2 emission scenario. *WATCH* Technical Report 49.

Gustavsson, J., Cederberg, C., Sonesson, U., van Otterdijk, R., & Meybeck, A. (2011). *Global food losses and food waste*. Rome, Italy: Food and Agriculture Organization of the United Nations.

GWP (Global Water Partnership). (2000). *Integrated water resources management*. Stockholm, Sweden: Global Water Partnership (Technical Advisory Committee Background Paper No. 4).

Haddeland, I., Clark, D. B., Franssen, W., Ludwig, F., Voß, F., Arnell, N. W., et al. (2011). Multimodel estimate of the global terrestrial water balance: Setup and first results. *Journal of Hydrometeorology, 12*, 869–884.

Hagemann, S., Chen, C., Haerter, J. O., Heinke, J., Gerten, D., & Piani, C. (2011). Impact of a statistical bias correction on the projected hydrological changes obtained from three GCMs and two hydrology models. *Journal of Hydrometeorology, 12*, 556–578.

Hajer, M. (2011). *The energetic society. In search of a governance philosophy for a clean economy*. The Hague, The Netherlands: PBL Netherlands Environmental Assessment Agency.

Hall, A. A., Rood, S. B., & Higgins, P. S. (2011). Resizing a river: A downscaled, seasonal flow regime promotes riparian restoration. *Restoration Ecology, 19*(3), 351–359.

Hansen, J. W., Challinor, A., Ines, A., Wheeler, T., & Moron, V. (2006). Translating climate forecasts into agricultural terms: Advances and challenges. *Climate Research, 33*, 27–41.

Headey, D., & Fan, S. (2010). *Reflections on the global food crisis. How did it happen? How has it hurt? And how can we prevent the next one?* Research Monograph, 165. Washington, DC: IFPRI.

Heilig, G. (1999). *China food: Can China feed itself*. Laxenburg, Austria: International Institute for Applied Systems Analysis.

Hirschman, A. O. (1975). Policymaking and policy analysis in Latin America—A return journey. *Policy Sciences, 6*(4), 385–402.

Hoekstra, A. Y., & Chapagain, A. K. (2008). *Globalization of water. Sharing the planet's freshwater resources*. Oxford, UK: Blackwell.

Hoekstra, A. Y., & Mekonnem, M. M. (2012). The water footprint of humanity. *Proceedings of the National Academy of Sciences of the United States of America, 109*(9), 3232–3237.

Hoff, H. (2011). *Understanding the nexus*. Stockholm, Sweden: Stockholm Environment Institute (Background Paper for The Bonn 2011 Conference: The Water, Energy and Food Security).

Holmberg, J., & Robert, K. H. (2000). Backcasting from non-overlapping sustainability principles: A framework for strategic planning. *International Journal of Sustainable Development and World Ecology, 74*, 291–308.

Howarth, R. D., Anderson, J., Cloern, C., Elfring, C., Hopkinson, B., Galloway, J. N., et al. (2002). Reactive nitrogen and the world: 200 years of change. *Ambio, 31*(2), 64–71.

Howarth, R. W., Santoro, R., & Ingraffea, A. (2011). Methane and the greenhouse-gas footprint of natural gas from shale formations. *Climate Change, 106*, 679–690.

IEA (International Energy Agency). (2008). *World energy outlook 2008*. Paris, France: IEA.

IFPRI (International Food Policy Institute). (2009). Climate change–Impact on agriculture and cost of adaptation – Washington DC, October.

IFPRI (International Food Policy Research Institute). (2010). *Food security, farming, and climate change to 2050: Scenarios, results, policy options*. Washington, DC: IFPRI.

IFPRI (International Food Policy Research Institute). (2012). *Finding the blue path for a sustainable economy*. Washington, DC: IFPRI (White Paper).

IIASA (International Institute for Applied Systems Analysis). (2012). *Global energy assessment: Toward a sustainable future*. Laxenburg, Austria: IIASA.

IIASA/FAO (International Institute for Applied Systems Analysis/Food and Agriculture Organization of the United Nations). (2012). *Global agro-ecological zones*. IIASA/FAO: Laxenburg, Austria/Rome, Italy (GAEZ v3.0).

IISD (International Institute for Sustainable Development). (2011). *Netley-Libau nutrient-bioenergy project*. Retrieved from http://www.iisd.org/pdf/2011/brochure_iisd_wic_netley_libau_2011.pdf

IISD (International Institute for Sustainable Development). (2011b). *Ecosystem approaches in integrated water resources management: A review of transboundary river basins*. Winnipeg, MB, Canada: International Institute for Sustainable Development.

IMF (International Monetary Fund). (2000). The world economy in the twentieth century: Striking developments and policy lessons. In *World Economic Outlook* (pp. 149–180). Washington, DC: IMF.

Inman, M. (2009, January 15). Where warming hits hard. *Nature Reports, Climate Change*. Retrieved from http://doi:10.1038/climate.2009.3

INRA & CIRAD (Institut National de la Recherche Agronomique & Centre International de la Recherche Agricole pour le Développement). (2010). *Agrimonde. Scénarios et défis pour nourrir le monde en*. Versailles, France: INRA/CIRAD (In French).

IPCC (Intergovernmental Panel on Climate Change). (2001). *Climate change 2001: Synthesis report, Summary for policymakers*. Third Assessment Report, IPCC.

IPCC (Intergovernmental Panel on Climate Change). (2007). *Climate change 2007: The physical science basis*. Fourth Assessment Report, IPCC.

IPCC (Intergovernmental Panel on Climate Change). (2012). *Managing the risks of extreme events and disasters to advance climate change adaptation*. A special report of working groups I and II, IPCC.

IWMI (International Water Management Institute). (2007). *Water for food, water for life. A comprehensive assessment of water management in agriculture.* London, UK/Colombo, Sri Lanka: Earthscan/IWMI.

Jackson, T. (2011). *Prosperity without growth. Economics for a finite planet.* London, UK: Earthscan.

Jacob, D., & van den Hurk, B. (2009). Climate change scenarios at global and local scales. In F. Ludwig, P. Kabat, H. van Schaik, & M. van der Valk (Eds.), *Climate change adaptation in the water sector* (pp. 23–34). London, UK: Earthscan.

Johnson, D. G. (1997). Agriculture and the wealth of nations. *The American Economic Review, 87,* 1–12.

Johnston, L., & Williamson, S. H. (2005). *The annual real and nominal GDP for the United States, 1789—Present.* Economic History Services. Retrieved from http://www.eh.net/hmit/gdp/

Kabat, P., Fresco, L. O., Stive, M. J., Veerman, C. P., van Alphen, J. S., Parmet, B. W., et al. (2009). Dutch coasts in transition. *Nature Geoscience, 2,* 450–452.

Kabat, P., van Vierssen, W., Veraart, J., Vellinga, P., & Aerts, J. (2005). Climate proofing the Netherlands. *Nature, 438,* 283–284.

Kaufmann, D., Kraay, A., & Mastruzzi, M. (2006). *Governance Matters V: aggregate and individual governance indicators for 1996–2005.* Washington, DC: World Bank.

Kenny, J. F., Barber, N. L., Hutson, S. S., Linsey, K. S., Lovelace, J. K., & Maupin, M. A. (2009). *Estimated use of water in the United States in 2005.* United States Geological Survey Circular 1344. Retrieved from http://pubs.usgs.gov/circ/1344/.

Kolbert, E. (2011). Enter the Anthropocene: Age of man. *National Geographic, 219,* 60–77.

Lannerstad, M. (2009). Water realities and development trajectories. In *Global and local agricultural production dynamics.* PhD Thesis. Linköping Studies in Arts and Science. No. 475. Linköping, Sweden: Linköping University Electronic Press.

Lapointe, T. M., Marcus, N., McGlathery, K., Sharpley, A., & Walker, D. (2000). Nutrient pollution of coastal rivers, bays, and seas. *Issues in Ecology, 7*(Fall).

Lindsey, R. (2009). NASA: Earth observatory. Retrieved from http://earthobservatory.nasa.gov/Features/EnergyBalance/

Lockwood, M., Davidson, J., Curtis, A., Stratford, E., & Griffith, R. (2008). Governance principles for natural resources management. *Society and Natural Resources, 23,* 1–16.

Lovins, A. B. (1976). Energy strategy: The road not taken? *Foreign Affairs, 55,* 63–96.

Lovins, A. B. (1977). *Soft energy paths: Toward a durable peace.* San Francisco: Friends of the Earth, International.

Ludwig, F., & Moench, M. (2009). The impacts of climate change on water. In F. Ludwig, P. Kabat, H. van Schaik, & M. van der Valk (Eds.), *Climate change adaptation in the water sector* (pp. 35–51). London, UK: Earthscan.

Luft, G. (2010). *Water crisis, energy crisis, vicious cycle.* Retrieved from http://www.huffingtonpost.com/gal-luft/water-crisis-energy-crisi_b_408518.html

Lundqvist, J. (2010). Producing more or wasting less. Bracing the food security challenge of unpredictable rainfall. In L. Martínez-Cortina, G. Garrido, & L. López-Gunn (Eds.), *Re-thinking water and food security: Fourth botín foundation water workshop.* London, UK: Taylor & Francis.

Lundqvist, J., de Fraiture, C., & Molden, D. (2008). Saving water: From field to fork—curbing losses and wastage in the food chain. In *SIWI policy brief.* Stockholm, Sweden: SIWI.

Lundqvist, J., & Falkenmark, M. (2010). Adaptation to rainfall variability and unpredictability. New dimensions of old challenges and opportunities. *International Journal of Water Resources Development, 26*(4), 597–614.

Lutz, W., & Samir, K. C. (2010). Dimensions of global population projections: What do we know about future trends and structures. *Philosophical Transactions of the Royal Society of London. Series B, Biological Sciences, 365,* 2779–2791.

Lutz, W., & Scherbov, S. (2008). *Exploratory extension of IIASA's world population projections: Scenarios to 2300.* Interim Report IR-08-022. Laxenburg, Austria: IIASA.

Lyons, B. (2012). *Primary energy and transportation fuels and the energy and water nexus: ten challenges.* Atlantic Council Energy and Environment Program Issue Brief. Washington, DC: Atlantic Council.

Mackay, H. (2003). Water policies and practices. In D. Reed & M. de Wit (Eds.), *Towards a just South Africa. The political economy of natural resource wealth* (pp. 49–83). Washington, DC/ Pretoria, South Africa: WWF Macroeconomics Programs Office/CSIR-Environmentek.

Madison, A. (1995). *Monitoring the world economy 1820–1992*. Paris, France: OECD.

Mark, B., & Selzer, G. (2003). Tropical glacier melt water contribution to stream discharge: A case study in the Cordillera Blanca, Peru. *Journal of Glaciology, 49*(165), 271–281.

Markel, D. (2005). *Monitoring and managing Lake Kinneret and its Watershed, Northern Israel: A response to environmental, anthropogenic and political constraints.* Retrieved from http:// agris.fao.org/agris-search/search/display.do?f=2006/QC/QC0601.xml;QC2005002208

McIntyre, N. E., Knowles-Yánez, K., & Hope, D. (2000). Urban ecology as an interdisciplinary field: Differences in the use of 'urban' between the social and natural sciences. *Urban Ecosystems, 4*, 5–24.

MDBA (Murray Darling Basin Authority). (2011). *The draft basin plan.* Canberra, ACT, Australia: MDBA.

MEA (Millennium Ecosystem Assessment). (2005). *Freshwater ecosystem services.* Washington, DC: Island Press.

Meadows, D. H., Meadows, D. L., Randers, J., & Behrens, W. W., III. (1972). *The limits to growth.* London, UK: East Island.

Means, E. G., III. (2012). Water 2050: Attributes of sustainable water supply development. In W. M. Grayman, D. P. Loucks, & L. Saito (Eds.), *Toward a sustainable water future: Visions for 2050.* Washington, DC: American Society of Civil Engineers.

Middelkoop, H., Daamen, K., Gellens, D., Grabs, W., Kwadijk, J., Lang, H., et al. (2001). Impact of climate change on hydrological regimes and water resources management in the Rhine Basin. *Climatic Change, 49*(1–2), 105–128.

Milly, P. C., Dune, K. A., & Vecchia, A. V. (2005). Global pattern trends in streamflow and water availability in a changing climate. *Nature, 438*, 347–350.

Moss, J., Wolf, G., Gladden, G., & Guttieriez, E. (2003). *Shifting paradigm: Towards a new economy of water for food and ecosystems.* Report of the African Pre-Conference on water for food and ecosystems Annex C-2.

MRC (Mekong River Commission) Secretariat. (2009). *Inception report: MRC SEA for hydropower on the Mekong mainstream.* Phnom Penh, Cambodia: MRC/International Center for Environmental Management.

Munthe, C. (2011). *The price of precaution and the ethics of risk.* New York: Springer.

Nellemann, C., MacDevette, M., Manders, T., Eikhout, B., Svihus, B., Prins, A. G., et al. (Eds.). (2009). *The environmental food crisis. The environment's role in averting future food crises.* Nairobi, Kenya: United Nations Environmental Programme.

Nelson, G. C., Rosegrant, M. W., Koo, J., Robertson, R., Sulser, T., Zhu, T., et al. (2009). *Climate change: Impact on agriculture and cost of adaptation.* Washington, DC: International Food Policy Research Institute.

Nelson, G. C., Rosegrant, M. W., Palazzo, A., Gray, I., Ingersoll, C., Robertson, R., et al. (2010). *Food security, farming, and climate change to 2050. Scenarios, results, policy options.* Washington, DC: International Food Policy Research Institute.

NETL (National Energy Technology Lab). (2010, September 13–17). CO_2 capture technology meeting.

Nicholson, S. (2005). On the question of the 'recovery' of the rains in the West African Sahel. *Journal of Arid Environments, 63*(3), 615–641.

North, D. C. (1994). Economic performance through time. *The American Economic Review, 84*(3), 359–368.

Norton, R. D. (2004). *Agricultural development policy—Concepts and experiences.* West Sussex, UK: John Wiley.

NRC (National Research Council). (2008). *Water implications of biofuels production in the United States.* Washington, DC: National Academy Press.

NRC (National Research Council). (2006). *Drinking water distribution systems: Assessing and reducing risks.* Washington, DC: National Academies Press.

NWC (National Water Commission). (2010). *Australian water markets report 2009–2010.* Canberra, ACT, Australia: NWC.

O'Grady, E. (2011, August 4). Heat waves pushes Texas power grid into red zone. *Reuters.*

O'Keeffe, J. (2009). Sustaining river ecosystems: Balancing use and protection. *Progress in Physical Geography, 33,* 339–357.

Oclay Ünver, I. H. (ed.). (1997). *Water resources development in a holistic socioeconomic context: The Turkish experience: Vol. 13, Issue 4 of International Journal of Water Resources Development.* Abingdon, UK.

Odum, E. P. (1997). *Ecology: A bridge between science and society.* Sunderland, MA: Sinauer.

OECD (Organisation for Economic Co-operation and Development). (2010). *Sustainable management of water resources in agriculture.* Paris, France: Organisation for Economic Co-operation and Development.

OECD (Organisation for Economic Co-operation and Development). (2011). Water governance in OECD countries. A multi-level approach. *OECD Studies on water.* Paris, France: OECD

OECD (Organisation for Economic Co-operation and Development). (2012). *Environment outlook to 2050: Freshwater chapter.* Paris, France: Organisation for Economic Co-operation and Development.

OECD-FAO (Organisation for Economic Co-operation and Development and Food and Agriculture Organization of the United Nations). (2012). Agricultural outlook 2012–2021. France: OECD and FAO. Retrieved from http://dx.doi.org/10.1787/agr_outlook-2012-eng

Oerlemans, J. (2005). Extracting a climate signal from 169 glacier records. *Science, 308*(5722), 675–677.

Oki, T., Valeo, C., & Heal, K. (Eds.). (2006). *Hydrology 2020: An integrating science to meet world water challenges.* Wallingford, UK: IAHS.

Owen, D. (2010, December 20, 27). The efficiency dilemma. *The New Yorker.*

Pacala, S., & Socolow, R. (2004). Stabilization wedges: Solving the climate problem for the next 50 years with current technologies. *Science, 305,* 968–972.

Palaniappan, M., Gleick, P. H., Allen, L., Cohen, M. J., Christian-Smith, J. and Smith, C. (ed. N. Ross) 2010. *Clearing the waters: A focus on water quality solutions.* Nairobi, Kenya: UNEP/ Pacific Institute. Retrieved from http://www.pacinst.org/reports/water_quality/clearing_the_waters.pdf

Parfitt, J., & Barthel, M. (2010). *Global food waste reduction: Priorities for a world in transition. Science Review* SR56. London, UK: Foresight, Government Office for Science.

Pate, R., Hightower, M., Cameron, C., & Einfeld, W. (2007). *Overview of energy-water interdependencies and the emerging energy demands on water resources.* Albuquerque, NM: Sandia National Laboratories.

Perrone, D., Murphy, J., & Hornberger, G. M. (2011). Gaining perspective on the water–energy nexus at the community scale. *Environmental Science & Technology, 45,* 4228–4234.

Phdungsilp, A. (2011). Futures studies' backcasting method used for strategic sustainable city planning. *Futures, 43,* 707–714.

Piani, C., Weedon, G. P., Best, M., Gomes, S. M., Viterbo, P., Hagemann, S., et al. (2010). Statistical bias correction of global simulated daily precipitation and temperature for the application of hydrological models. *Journal of Hydrology, 395*(3–4), 199–215.

Pilgrim, N., Roche, B., Kalbermatten, J., Revels, C., & Kariuki, M. (2007). *Principles of town water supply and sanitation.* Washington, DC: Word Bank (Water Working Note No. 13).

Pittock, J., & Connell, D. (2010). Australia demonstrates the planet's future: Water and climate in the Murray-Darling Basin. *International Journal of Water Resources Development, 26*(4), 561–577.

Postel, S. (1999). *Pillar of sand. Can the irrigation miracle last?* New York: Worldwatch Institute.

Power, S., Sadler, B., & Nicholls, N. (2005). The influence of climate science on water management in Western Australia. *Bulletin of the American Meteorological Society, 86*(6), 839–844.

Preston, B., Smith, T., Brooke, C., Gorddard, R., Measham, T., Withycombe, G., et al. (2008). *Mapping climate change vulnerability in the Sydney Coastal Councils Group.* Hobart, TAS, Australia: CSIRO Marine and Atmospheric Research.

Pyper, C., & ClimateWire. (2011). Electricity generation "burning" rivers of drought-scorched southeast. *Scientific American*. Retrieved from http://www.scientificamerican.com/article. cfm?id=electricity-generation-buring-rivers-drought-southeast

Quist, J. (2007, May10). Backcasting for sustainable futures and system innovations. TiSD–Colloquium advanced course. Delft, The Netherlands: Delft University of Technology.

Raskin, P., Gallopin, G., Gutman, P., Hammond, A. and Swart, R. 1998. Bending the curve: Toward global sustainability. *Polestar series report No. 8*. Boston: Stockholm Environment Institute.

Raskin, P., Gleick, P., Kirshen, P., Kirshen, G., & Strzepek, K. (1997). *Water futures: Assessment of long-range patterns and problems*. Boston: Stockholm Environment Institute.

Reanalysis.org. (2012). Reanalysis intercomparison and observations. Retrieved from http://reanalyses.org

Rees, W. E. (2003). Understanding urban ecosystems: An ecologic economics perspective. In A. R. Berkowitz, C. H. Nilon, & K. S. Kollweg (Eds.), *Understanding urban ecosystem: A new frontier for sciences and education* (pp. 115–136). New York: Springer.

Reilly, M., Willenbockel, D. (2010). Managing uncertainty: A review of food system scenario analysis and modelling. *Philosophical Transactions of the Royal Society, 365*, 3049–3063. (doi:10.1098/rstb.2010.0141).

Reisner, M. (1986). *Cadillac Desert: The American west and its disappearing water*. New York: Viking Penguin.

Richter, B. D. (2009). Re-thinking environmental flows: From allocations and reserves to sustainability boundaries. *River Research and Applications, 25*, 1–12.

Robinson, J. B. (1982). Energy backcasting: A proposed method of policy analysis. *Energy Policy, 10*, 337–344.

Rogers, P., & Hall, A. W. (2003). *Effective water governance*. Stockholm, Sweden: Global Water Partnership (Technical Advisory Committee Background Papers No. 7).

Rönnbäck, P., Bryson, I., & Kautsky, N. (2002). Coastal aquaculture development in eastern Africa and the western Indian ocean: Prospects and problems for food security and local economies. *Ambio, 31*, 537–542.

Rood, S. B., Gourley, C. R., Ammon, E. M., Heki, L. G., Klotz, J. R., Morrison, M. L., et al. (2003). Flows for floodplain forests: A successful riparian restoration. *BioScience, 53*, 647–656.

Rosegrant, M. W., Cai, X., & Cline, S. A. (2002). *World water and food to 2025: Dealing with scarcity*. Washington, DC: International Food Policy Research Institute.

Sachs, J. (2008, October 21). Amid the rubble of global finance, a blueprint for Bretton Woods II. *The Guardian*.

Schumpeter, J. (2011, June 8). Energy statistics. The world gets back to burning. *The Economist*.

Schwartz, P. (1991). *The art of the long view*. New York: Currency/Doubleday Press.

Scudder, T. (1993). *The IUCN review of the Southern Okavango Integrated Water Development Project*. Gland, Switzerland: IUCN.

Seckler, D. (1996). *The new era of water resources management: from 'dry' to 'wet' water savings*. Research Report 1. Colombo, Sri Lanka: International Irrigation Management Institute

Seckler, D., Amarasinghe, U., Molden, D., de Silva, R., & Barker, R. (1998). *World water demand and supply, 1990 to 2025: Scenarios and issues*. Colombo, Sri Lanka: International Water Management Institute (Research Report 19).

Shackleton, C., Shackleton, S., Gambiza, J., Nel, E., Rowntree, K., & Urquhart, P. (2008). *Links between ecosystem services and poverty alleviation: Situation analysis for arid and semi-arid lands in southern Africa*. Consortium on Ecosystems and Poverty in Sub-Saharan Africa (CEPSA).

Shah, T. (2009). *Taming the anarchy: Groundwater governance in South Asia*. Washington, DC: Resources for the Future Press.

Shah, A. (2010). Poverty facts and stats. Retrieved from http://www.globalissues.org/article/26/poverty-facts-and-stats

Shah, T., Molden, D., Sathivadivel, R., & Seckler, D. (2000). *The global groundwater situation: Overview of opportunities and challenges*. Colombo, Sri Lanka: IWMI.

Sharma, D., Das Gupta, A., & Babel, M. S. (2007). Spatial disaggregation of bias-corrected GCM precipitation for improved hydrologic simulation: Ping River Basin, Thailand. *Hydrology and Earth System Sciences, 11*, 1373–1390.

Shehabi, A., Stokes, J. R., & Horvath, A. (2012). Energy and air emission implications of a decentralized wastewater system. *Environmental Research Letters, 7*, 024007.

Shiklomanov, I. A. (1993). World fresh water resources. In P. H. Gleick (Ed.), *Water in crisis: A guide to the world's fresh water resources* (pp. 13–24). New York: Oxford University Press.

Shiklomanov, I. A. (1998). *Assessment of water resources and water availability in the world.* St. Petersburg, Russia: State Hydrological Institute (Report for the Comprehensive Assessment of the Freshwater Resources of the World, United Nations. Data Archive on CD-ROM).

Sills, B. (2011). Fossil fuel subsidies six times more than renewable energy. Bloomberg. Retrieved from http://www.bloomberg.com/news/2011-11-09/fossil-fuels-got-more-aid-than-clean-energy-iea.html

Skov Andersen, L. (2011). *China—Three gorges and a remedy.* (Stockholm Water Front. No. 2). Retrieved from http://www.siwi.org/documents/Resources/Water_Front_Articles/2011/WF_2_2011_China.pdf

Smakhtin, V. (2008). Basin closure and environmental flow requirements. *International Journal of Water Resources Development, 24*(2), 227–233.

Smith, I. (2004). An assessment of recent trends in Australian rainfall. *Australian Meteorological Magazine, 53*, 163–173.

Sood, A., Chartres, C. J., Lundqvist, J., & Ait Kadi, M. (2013). *Global water demand scenarios 2010–2050.* Stockholm, Sweden: IWMI, Colombo and SIWI.

Stahre, P. (2008). *Sustainable urban drainage: Blue-green fingerprints in the city of Malmö, Sweden.* Malmö, Sweden: VA SYD. Retrieved from http://www.vasyd.se/fingerprints

Starkl, M., Parkinson, P., Narayana, D., & Flamand, P. (2012). Small is beautiful but is large more economical? Fresh views on decentralised versus centralised wastewater management. *Water, 21*, 45–47.

Steffen, W., Crutzen, P. J., & McNeill, J. R. (2007). The Anthropocene: Are humans now overwhelming the great forces of nature? *Ambio, 36*(8), 614–621.

Steffen, W., Persson, A., Deutsch, L., Zalasiewicz, J., Williams, M., Richardson, K., et al. (2011). The Anthropocene: From global change to planetary stewardship. *Ambio, 40*(October), 739–761.

Stern, N. (2007). *The economics of climate change: The Stern review.* Cambridge, UK: Cambridge University Press.

SWITCH (Sustainable Water Management in the City of the Future). (2011). *Findings from the SWITCH Project 2006–2011.* Paris, France: UNESCO-IHE Institute for Water Education.

TEEB (The Economics of Ecosystems and Biodiversity). (2009, September). the economics of ecosystems and biodiversity of climate issues update. Switzerland. Retrieved from http://www.teebweb.org/publications/teeb-study-reports/foundations/.

The Economist. (2011, May 26, 2011). A man made world: The Anthropocene. *The Economist.* London, UK. Retrieved from http://www.economist.com/node/18741749

The Natural Step (2012) Backcasting. Retrieved from http://www.naturalstep.org/backcasting

Torero, M. (2011). *Riding the rollercoaster.* Washington, DC: International Food Policy Institute (IFPRI Global Food Policy Report).

Tsegaye, S., Eckhart, J., & Vairavamoorthy, K. (2011). Urban water management in the cities of the future: Emerging areas in developing countries. In J. Lundqvist (Ed.), *On the water front: Selections from the 2011 world water week in Stockholm* (pp. 42–47). Stockholm, Sweden: SIWI.

Tucci, C. E. M. (2009). *Integrated urban water management in large cities: A practical tool for assessing key water management issues in the large cities of the developing world.* Washington, DC: World Bank (Draft Paper Prepared for World Bank).

Tucci, C. E. M., Goldenfum, J. A., &Parkinson J. N. (eds.). (2010). *Integrated urban water management: humid tropics. UNESCO-IHP: Urban Water Series.* Boca Raton, FL: CRC Press.

UN (United Nations). (1992). *Report of the United Nations conference on environment and development.* New York: United Nations.

UN (United Nations). 1998. Strategic approaches to freshwater management. *6th Session of the commission on sustainable development*. New York: United Nations.

UN (United Nations). (2004). *World population to 2300*. New York: UN Department of Economic and Social Affairs.

UN GA (General Assembly of the United Nations). *United Nations General Assembly Resolution A/RES/37/7*. Retrieved from http://www.un.org/documents/ga/res/37/a37r007.htm

UN HABITAT. (2011). *The state of the world's cities 2010/2011—Cities for All: bridging the urban divide*. London, UK: Earthscan.

UN WWAP (United Nations World Water Assessment Programme). (2006). The United Nations world water development report 2: *Water a shared responsibility*. Paris, France: UNESCO.

UN WWAP (United Nations World Water Assessment Programme). (2009). The United Nations World Water Development. Report 3. *Water in a changing world*. Paris, France/London, UK: UNESCO/Earthscan.

UN WWAP (United Nations World Water Assessment Programme). (2012). The United Nations World Water Development, Report 4. *Managing water under risk and uncertainty*. Paris, France: UNESCO.

UN, WWAP (United Nations World Water Assessment Programme). (2011). *World water scenarios to 2050, exploring alternative futures of the world's water and its use to 2050*. Paris, France: UNESCO.

UNDESA (United Nations Department of Economic and Social Affairs). (2009). *The challenges of adapting to a warmer planet for urban growth and development*. New York: UNDESA. (UN-DESA Policy Brief No. 25).

UNDP (United Nations Development Programme). (2011). Human development report. In *Sustainability and equity: A better future for all*. New York: UNDP.

UNEP (United Nations Environment Programme). (2002). *Global environmental outlook 3: Past, present and future Perspectives*. London, UK/Nairobi: Earthscan/UNEP.

UNEP (United Nations Environment Programme). (2006). *Marine and coastal ecosystems and human well-being: A synthesis report based on the findings of the Millennium Ecosystem Assessment*. Nairobi, Kenya: UNEP.

UNEP (United Nations Environment Programme). (2009). *Towards sustainable production and sustainable use of resources: Assessing biofuels*. Paris, France: UNEP (Produced by the International Panel for Sustainable Resource Management, Division of Technology Industry and Economics, France).

UNEP (United Nations Environment Programme). (2012). *UN global environmental monitoring system (GEMS)*. Canada. Retrieved from http://www.gemswater.org.

UNEP/RIVM (United Nations Environment Programme and Rijksinstituut voor Volksgezondheid en Milieu). (2004). *The GEO-3 scenarios 2002–2032. Quantification and Analysis of Environmental Impacts*, ed. J. Potting and J. Bakkes. Nairobi, Kenya: PBL Netherlands Environmental Assessment Agency.

UNESCO (United Nations Educational, Scientific and Cultural Organization). (2003). Overview. In C. I. Dooge, J. Delli Priscoli, & M. R. Llamas (Eds.), *Series on water and ethics, essay 1*. Paris, France: UNESCO.

UNFPA (United National Population Fund). (2011). *The state of world population*. New York: UNFPA.

UNFPA (United Nations Population Fund). (2007). *State of world population 2007: Unleashing the potential of urban growth*. New York: UNFPA.

UNFPA (United Nations Population Fund). (2010). *World population prospects, the 2010 revision*. New York: UNFPA.

USDA, FAS (United States Department of Agriculture Foreign Agricultural Service). (2011). Production, supply and distribution online. Retrieved from http://www.fas.usda.gov/psdonline

USDOE (United States Department of Energy). (2006). *Energy demand on water resources*. Washington, DC: USA (Report to Congress on the Interdependence of Energy and Water).

USDOE (United States Department of Energy). (2008). *Water requirements for existing and emerging thermoelectric plant technologies*. Morgantown WV: National Energy Technology Laboratory.

USDOE (United States Department of Energy). (2011). *Water heating*. Retrieved from http://www.energysavers.gov/your_home/water_heating/index.cfm/mytopic=12760

USGS (United States Geological Survey). (2012). National streamflow information program. Reston, VA. Retrieved from http://water.usgs.gov/nsip/history1.html

Usman, M. T., & Reason, C. J. (2004). Dry spell frequencies and their variability over southern Africa. *Climate Research, 26*, 199–211.

van den Hurk, B., & Jacob, D. (2009). The art of predicting climate variability and change. In F. Ludwig, P. Kabat, H. van Schaik, & M. van der Valk (Eds.), *Climate change adaptation in the water sector* (pp. 9–22). London, UK: Earthscan.

van der Steen, P. (2006, January). *Integrated urban water management: Towards sustainability*. Paper presented at the first SWITCH scientific meeting, University of Birmingham, Birmingham, UK.

van der Veer, J. (2010). *Water—A critical enabler to produce energy*. Amsterdam, The Netherlands: Royal Dutch Shell.

van Vliet, M. T. H., Franssen, W. H. P., Yearsley, J. R., Ludwig, F., Haddeland, I., Lettenmaier, D. P., et al. (2013). Global river discharge and water temperature under climate change. *Global Environmental Change, 23*(2), 450–464.

van Vliet, M. T., Yearsley, J. R., Ludwig, F., Vögele, S., Lettenmaier, D. P., & Kabat, P. (2012, June 4). Vulnerability of US and European electricity supply to climate change. *Nature Climate Change*. Retrieved from http://www.doi:10.1038/nclimate1546.

Vlachos, E., & Braga, B. P. F. (2011). The challenge of urban water management. In C. Maksimovic & J. Tejada-Guibert (Eds.), *Frontiers in urban water management: Deadlock or hope* (pp. 1–36). Paris, France: UNESCO.

Voinov, A., & Cardwell, H. (2009). The energy–water nexus: Why should we care? *Contemporary Water Research & Education, 143*, 17–29.

von Braun, J. (2008). The rise in food and agricultural prices—Implications for Morocco. Rabat, Morocco: IRES (Royal Institute of Strategic Studies).

Vörösmarty, C. J., Green, P., Salisbury, J., & Lammers, R. B. (2000). Global water resources: Vulnerability from climate change and population growth. *Science, 289*(5477), 284–288.

Vörösmarty, C. J., McIntyre, P. B., Gessner, M. O., Dudgeon, D., Prusevich, A., Green, P., et al. (2010). Global threats to human water security and river biodiversity. *Nature, 467*, 555–561.

Wada, Y., van Beek, L. P. H., Bierkens, M. F. P. (2012). Nonsustainable groundwater sustaining irrigation: A global assessment. *Water Resources Development, 48*(6), 18. W00L06, doi:10.1029/2011WR010562.

Wald, M. L. (2012, July 17). So, how hot was it? *New York Times*.

Walton, B. (2010). *Low water may halt Hoover Dam's power*. Traverse City, MI: Circle of Blue.

Wang, X.-Y., Tao, F., Xiao, D., Lee, H., Deen, J., Gong, J., et al. (2006). Trend and disease burden of bacillary dysentery in China (1991–2000). *Bulletin of the World Health Organization, 84*, 561–568.

Wani, S. P., Rockström, J., & Venkateswarlu, B. (2011). New paradigm to unlock the potential of rainfed agriculture in the semiarid tropics. In R. Lal & B. A. Steward (Eds.), *World soil resources and food security* (pp. 419–470). Baton Rouge, FL: CRC Press.

WCD (World Commission on Dams). (2000). *Dams and development. A new framework for decision-makers*. London, UK: Earthscan.

WCED (World Commission on Environment and Development.). (1987). *Our common future*. Oxford, UK: Oxford University Press.

Weaver, P., Jansen, L., van Grootveld, G., van Spiegel, E., & Vergragt, P. (2000). *Sustainable technology development*. Sheffield, UK: Greenleaf.

Webber, M. E. (2008). Catch-22: Water vs. Energy. *Scientific American, 18*, 34–41.

Webster, P. J., Holland, G., Curry, J., & Chang, H. R. (2005). Changes in tropical cyclone number, duration, and intensity in a warming environment. *Science, 309*(5742), 1844–1846.

Weedon, G. P., Gomes, S., Viterbo, P., Österle, H., Adam, J. C., Bellouin, N., et al. (2010). The WATCH forcing data 1958–2001: A meteorological forcing dataset for land surface and hydrological models. WATCH. Retrieved from http://www.eu-watch.org/publications/technical-reports

WEF (World Economic Forum). (2009). *Thirsty energy: Water and energy in the 21st century.* Geneva, Switzerland: WEF.

WEF (World Economic Forum). (2011). *Global risks. Executive summary* (6th ed.). Geneva, Switzerland: WEF.

WEFWI (World Economic Forum Water Initiative). (2011). *Water security. The water-food-energy-climate nexus.* Washington, DC: Island Press.

White, G. F. (1945). *Human adjustment to floods: A geographical approach to the flood problem in the United States.* Chicago, IL: Department of Geography, University of Chicago (Research Paper No. 29).

WHO (World Health Organization). (2002). *World health report: Reducing risks, promoting healthy life.* Geneva, Switzerland: WHO.

WHO-UNDP (World Health Organization and United Nations Development Programme). (2001). Environment and people's health in China. Retrieved from www.un.org/esa/sustdev/publications/trends2006/endnotes.pdf

Wilde, K. (2010). Baboons in pinstripes: The inevitable target of an 'appropriate' economics. *World Future Review., 2,* 24–40.

Wilkinson, M. E., Quinn, P. F., Benson, I., & Welton, P. (2010). Runoff management: mitigation measures for disconnecting flow pathways in the Belford Burn catchment to reduce flood risk. *British Hydrological Society Third International Symposium,* Managing Consequences of a Changing Global Environment, Newcastle, UK.

Williams, M., Zalasiewicz, J., Haywood, A., & Ellis, M. (2011). The Anthropocene: A new epoch of geological time. *Philosophical Transactions of the Royal Society, 369,* 835–1111.

Wilson, M. A., & Carpenter, S. R. (1999). Economic valuation of freshwater ecosystem services in the united states: 1971–1997. *Ecological Applications, 9*(3), 772–783.

Wittfogel, K. A. (1956). *The hydraulic civilization: Man's role in changing the earth.* Chicago, IL: University of Chicago Press.

Wolff, G., & Gleick, P. H. (2002). The soft path for water. In P. H. Gleick (Ed.), *The world's water 2002–2003: The biennial report on freshwater resources* (pp. 1–32). Washington, DC: Island Press.

Woodside, C. (2011). Energy efficiency and 'the rebound effect'. Retrieved from http://www.yale-climatemediaforum.org/2011/03/energy-efficiency-and-the-rebound-effect

World Bank. (2009). *World Development Report 2009. Reshaping economic geography.* Washington, DC: World Bank.

World Bank. (2011). *Global monitoring report 2011: Improving the odds of achieving the MDGs—Heterogeneity, gaps and challenges: Overview.* Washington, DC: World Bank.

World Bank. (2012). *Global economic prospects 2012. Uncertainties and vulnerabilities.* Retrieved from http://go.worldbank.org/WI8LCZ6PT.0

World Energy Council. (2001). *Living in one world.* London, UK: World Energy Council.

World Energy Council. (2010). *Water for energy.* London, UK: World Energy Council.

Wouters, P., Hu, D., Zhang, J., Tarlock, A. D., & Andrews-Speed, P. (2004). The new development of water law in China. *University of Denver Water Law Review, 7*(2), 243–308.

WWAP (World Water Assessment Programme). (2009). *World water development report 3.* Paris, France: UNESCO.

WWAP (World Water Assessment Programme). (2012). *The United Nations world water development report 4: Managing water under risk and uncertainty.* Paris, France: UNESCO.

WWC (World Water Council). (2000). *Final report: Second world water forum (The Hague).* Marseille, France: WWC.

WWF (World Wildlife Fund). (2011). *The energy report: 100 % renewable energy by 2050.* Gland, Switzerland: WWF.

Xie, J., Liebenthal, A., Warford, J. J., et al. (2009). *Addressing China's water scarcity: Recommendations for selected water resource management issues.* Washington, DC: World Bank.

Yoshimoto, T., & Suetsugi, T. (1990). Comprehensive flood disaster prevention measures in Japan. In H. Massing (Ed.), *Hydrological processes and water management in urban areas* (pp. 175–183). Wallingford, UK: International Association of Hydrological Sciences.

Young, G. J., Dooge, J. C. I., & Rodda, J. C. (1994). *Global water resource issues*. Cambridge, UK: Cambridge University Press.

Yuan, Z., & Tolb, R. (2004). *Evaluating the costs of desalination and water transport*. Hamburg, Germany: University of Hamburg (Working Paper 41).

Zalasiewicz, J., Williams, M., Fortey, R., Smith, A., Barry, T. L., Coe, A. L., et al. (2011). Stratigraphy of the Anthropocene. *Philosophical transactions. Series A, Mathematical, Physical, and Engineering Sciences, 309*, 1036–1105.

Zhang, J., Mauzerall, D. L., Zhu, T., Liang, S., Ezzati, M., & Remais, J. V. (2010). Environmental health in China: Progress towards clean air and safe water. *The Lancet, 375*, 1110–1119.

Zhou, B., & Wang, Q. (2009). Strategy adjustment in flood control, disaster reduction and flood risk management. *Water Conservancy Science and Technological and Economy, 15*(4), 319–320.

Biographic Notes of the Members of the Gulbenkian Think Tank on Water and the Future of Humanity

Benedito Braga is a Professor of Civil and Environmental Engineering at University of São Paulo, Brazil. He holds a M.S. in Hydrology and Ph.D. in Water Resources from Stanford University. Prof. Braga research interests include urban water management, multiobjective modeling of water resources systems, and water policy development. He is the author of more than 200 articles and 25 books and book chapters on water management. He served on the board of Directors of the Brazilian National Water Agency from 2001 to 2009. He was President of the International Water Resources Association in 1998–2000, President of the Intergovernmental Council of IHP-UNESCO (2008–2009), and President of the International Committee of the sixth World Water Forum in Marseille in 2012. He is now President of the World Water Council. Prof. Braga is recipient of the 2002 Crystal Drop Award, given by the International Water Resources Association (IWRA) in recognition for his life time achievements in the area of water resources management. In 2011, he became Diplomate of the American Academy of Water Resources (ASCE-EWRI) for his eminence in water resources.

Colin Chartres has a Ph.D. on soil development from the University of Reading (UK). He is the former (2007–2012) Director General of the International Water Management Institute (IWMI). IWMI's vision is *Water for a Food Secure World* and involves solving water scarcity via increasing water productivity, reducing poverty, and sustainable natural resource management. He played a leading role in alerting the world to an emerging water crisis that will impact all water users and food security in many developing countries. Prior to joining IWMI he was Chief Science Advisor to Australia's National Water Commission. There, his role included developing a national water information system, creating a national groundwater action plan, and using scientific evidence to inform water policy options. Previously, he held senior research and research management positions with CSIRO, the Bureau of Rural Science and Geoscience Australia, and has also worked in academia and the private sector. He has published over 100 journal articles, technical papers, and book chapters on natural resources management and is the senior coauthor of the book "Out of Water" published in 2010.

Gulbenkian Think Tank on Water and the Future of Humanity, *Water and the Future of Humanity: Revisiting Water Security*, DOI 10.1007/978-3-319-01457-9, © Calouste Gulbenkian Foundation 2014

William J. Cosgrove received a B.Eng. and M.Eng. in Sanitary Engineering from McGill University and an honorary Doctorate of Science from the University's Faculty of Agricultural and Environmental Sciences. He has followed other graduate courses in economics, management, and cross-cultural studies. He is President of Ecoconsult, Inc. and Senior Research Scholar at the International Institute for Applied Systems Analysis and Director of the Water Futures and Solutions Initiative. He is a former Vice President of the World Bank, past President of the World Water Council, and served as Chairman of the International Steering Committee of the Dialogue on Water and Climate. He wrote the terms of reference for, and was a founding member of, the Global Water Partnership. From 2005 to 2008, he was President of the Bureau d'audiences publiques sur l'environnement in Québec. He was Content Coordinator for the third edition of the United Nations World Water Development Report and Senior Adviser for the fourth edition. His most recent publication is *The Dynamics of Global Water Futures: Driving Forces 2011–2050* (with Catherine E. Cosgrove, UNESCO, Paris, 2012).

Luis Veiga da Cunha has a Ph.D. in Civil Engineering from the Technical University of Lisbon. He is a Professor in the Department of Environmental Sciences and Engineering, of Universidade Nova de Lisboa. He is well known for his work nationally and internationally as Scientist, Professor, and Consultant in the fields of Hydrology, Water Resources Planning and Management, Water Policies, and Global Change and Water. He is a member of the Portuguese National Water Council and of the National Environment and Sustainable Development Council. He was Head of the Water Resources Division of the National Laboratory of Civil Engineering, Lisbon (1963–1983); Administrator of the Scientific and Environmental Affairs Division of NATO, Brussels (1983–1999); Member of IPCC (1989–2001); and Lead author of the Chapter on Hydrology and Water Resources of the Third Assessment Report of IPCC (1999–2001). He was founder and the first President of the Portuguese Water Resources Association (1977–1978). He is a member of the Portuguese Academy of Sciences, the Portuguese Academy of Engineering and the French Water Academy, being the first non-French member of this Academy. He was the Minister of Education of Portugal (1979–1980).

Peter H. Gleick is cofounder and president of the Pacific Institute in California. His research and writing address the critical connections between water and human health, hydrologic impacts of climate change, sustainable water use, privatization and globalization, and international conflicts over water resources. He is an internationally recognized water expert and was named a MacArthur Fellow in 2003 for his work. He was dubbed a "visionary on the environment" by the BBC. In 1999, he was elected an Academician of the International Water Academy, in Oslo, Norway and in 2006 he was elected to the US National Academy of Sciences. He received a B.S. from Yale University and M.S. and Ph.D. from the University of California, Berkeley. He serves on the boards of numerous journals and organizations, and is author of many scientific papers and nine books, including *Bottled and Sold: The Story Behind Our Obsession with Bottled Water* and the biennial report, *The World's*

Water (Island Press, Washington, DC), and a *Twenty-First Century US Water Policy* (Oxford University Press).

Pavel Kabat has a Ph.D. in Hydrology, Water Resources, and Amelioration. His fields of expertise are climate hydrology and water cycle, water resources and climate, global change, land–atmosphere interactions and (terrestrial) biochemical feedbacks (measurements and modeling), and climate system and climate change. He is the Director and CEO of the International Institute for Applied Systems Analysis—IIASA, Austria. He is full Professor and chair holder on Earth System Sciences and Climate Studies at the Wageningen University. He is Chair of the Board of the Wageningen Climate Centre, Science Director Council Chair of the Dutch National Climate Research Program, and Chair and Director of the Royal Dutch Academy of Sciences Institute for Integrated Research on Wadden Sea Region. He is Cochair, International Scientific Steering Committee of IGBP/ILEAPS program. He is a former member of the Delta Committee, former Chair International Scientific Steering Committee IGBP BAHC program, former Chair International Science Panel of GEWEX-ISLSCP/WCRP, and Lead Author IPCC fourth Assessment Report (2004–2007).

Mohamed Ait Kadi has a Ph.D. in Irrigation Engineering from Utah State University, USA and a Doctorate of Sciences in Agronomy. He is presently President of the General Council of Agricultural Development, a high level policy think tank of the Ministry of Agriculture and Fishery in Morocco. Previously as Director General of the Irrigation Department, he was in charge of the development and implementation of Morocco's National Irrigation Program. He was also a key player in the reform of Morocco's water sector. He chairs the Technical Advisory Committee of the Global Water Partnership. He was Governor and founding member of the World Water Council and President of the organizing committee of the first World Water Forum (Marrakech 1997). He is a member of Hassan II Academy of Sciences and Technology and member of the Board of the Consortium of CGIAR. He is a Professor at the Institute of Agronomy and Veterinary Medicine Hassan II in Rabat. He is the author of numerous publications in the fields of agriculture and rural development, irrigation, and water management.

Daniel P. Loucks is a Professor in the School of Civil and Environmental Engineering at Cornell University, specializing in the application of environmental engineering, economic theory, ecology, and systems analysis methods to environmental and regional water resources management issues. Over the past decades, he has held appointments at other universities in the USA, Europe, and Australia, at the World Bank, at the International Institute for Applied Systems Analysis in Austria and in various UN agencies. He has served on various scientific committees and boards of professional organizations including those of NATO, the US Academies, the National Research Council, and the US Army Corps of Engineers. He has served in private, government, and international organizations in the USA as well as Asia, Australia, Eastern and Western Europe, the Middle East, Africa, and Latin America. He has authored and coauthored many papers and books, including two widely used textbooks in water resources systems planning and management. He was elected a distinguished member of the American Society of Civil Engineers, an Honorary Diplomate of the American Academy of Water Resources, and a member of the US National Academy of Engineering.

Jan Lundqvist has a Ph.D. in Human Geography, University of Gothenburg (1975). He has served as Senior Lecturer and docent (Research Professor) at Universities of Bergen and Oslo, Norway (1973–1980), Professor at the Department of Water and Environmental Studies, Linköping University (1980–2009). He is a member of the Royal Academy of Agriculture and Forestry, Sweden. Currently, he is Senior Scientific Advisor at Stockholm International Water Institute. His research and consultancies deal with water and climate, water and energy for food security, river basin dynamics, and, generally, water in development. His current focus is on climate variability and unpredictability, food supply chain dynamics, and interdependence between freshwater and terrestrial systems and coastal zone. He has worked in Africa, South Asia, South East Asia, Middle East, and Scandinavia. He has contributed to the literature on water resources management and policy for socioeconomic development in global and regional contexts and held interviews in leading media: BBC, CNBC, Canadian TV, Financial Times, Xinhua News, Xiamen TV, and Scandinavian media.

Sunita Narain is the Director General of the Center for Science and Environment, New Delhi, where she worked since 1982. She is a member of the Society for Environmental Communications and publisher of the magazine "Down To Earth." A writer and environmentalist, she conducts research with forensic rigor and passion, so that knowledge can lead to change. In 2005, 2008, and 2009, she was cited by US journal "Foreign Policy" as one of the world's 100 public intellectuals. In 2005, she was awarded the Padma Shri by the Indian government. In 2005 CSE, under her direction, received the World Water Prize for work on rainwater harvesting and for its policy influence in building paradigms for community-based water management. In 2005, she chaired the Tiger Task Force at the direction of the Prime Minister. She advocated solutions to build a coexistence agenda with local communities so that benefits of conservation could be shared and the future secured. She is a member of the Prime Minister's Council for Climate Change and a member of the National Ganga River Basin Authority, chaired by the Prime Minister.

Jun Xia was the President of International Water Resources Association (2010–2012) and a Governor of *World Water Council.* Since 1987, he has lead water research managing and consulting jobs in China and internationally. He also had several leading roles in international water programmes and projects, such as Cochair of the Water Programme of the *Inter-Academy Council,* Director of *Asia Scientific Network Office for Global Water System Projec*t, Cochair of *Australia–China Center on Water Resources Research.* He is a member of Scientific Steering Committee for *Global Water System Project.* In China, he is a Chair Professor of Hydrology and Water Resources, State Key Laboratory of Water Resources and Hydropower Engineering, Dean at the Research Institute for Water Security (RIWS) Wuhan

University, and Distinguished Leading Professor, Key Laboratory of Water Cycle and Related Land Surface Processes, the Chinese Academy of Sciences (CAS). His major research fields relate to the complexity of water systems, hydrological processes and changing environment at river basin scale, climate change impact on water security, and water sustainable use and governance.

CPSIA information can be obtained
at www.ICGtesting.com
Printed in the USA
LVHW08*1927201018
594272LV00002B/22/P